# Organometallics
# in Organic Synthesis

# Chapman and Hall Chemistry Textbook Series

CONSULTING EDITORS

R. P. Bell, M.A., Hon. LLD., F.R.S., Professor of Chemistry at the University of Stirling

N. N. Greenwood, Ph.D., Sc.D., Professor of Inorganic and Structural Chemistry at the University of Leeds

R. O. C. Norman, M.A., D.Sc., Professor of Chemistry at the University of York

OTHER TITLES IN THE SERIES

*Symmetry in Molecules* J. M. Hollas

*N.M.R. and Chemistry* J. W. Akitt

*Introduction to Molecular Photochemistry* C. H. J. Wells

*The Chemistry of the Non-Metals* P. Powell and P. Timms

*Principles and Applications of Electrochemistry* D. R. Crow

*Pericyclic Reactions* G. B. Gill and M. R. Willis

# Organometallics in Organic Synthesis

## J. M. Swan

Pro-Vice-Chancellor,
Monash University

## D. St.C. Black

Senior Lecturer in Organic Chemistry,
Monash University

CHAPMAN AND HALL

LONDON

*First published 1974*
*by Chapman and Hall Ltd*
*11 New Fetter Lane, London EC4P 4FF*

© *1974 J. M. Swan, D. St.C. Black*

*Printed in Great Britain by*
*William Clowes & Sons Limited*
*London, Colchester and Beccles*

SBN 412 10870 4

Distributed in the U.S.A.
by Halsted Press, a Division
of John Wiley & Sons, Inc., New York

# Preface

This is a book about organic chemistry, especially the synthesis of organic compounds. However, we are not concerned with the whole field of synthesis but rather with those reactions where the starting material or an intermediate is a substance having a carbon–metal bond. Synthesis involving organometallic compounds or intermediates is a large, diverse, and important field, which is growing rapidly at the present time.

Organometallic chemistry had its origins and its first period of rapid growth in relation to organic synthesis. In particular, the success of the Grignard reagents stimulated chemists to explore the organic derivatives of a wide range of metals. In the second phase of growth, starting perhaps with ferrocene, much progress was made in theoretical studies of carbon–metal bonding, in the synthesis of organometallic compounds as such, and in detailed investigations of the physical and chemical properties of these substances. This second phase has left its mark on textbooks where mostly the subject is treated in terms of the nature of the metal atom concerned and its place in the Periodic Table rather than in terms of the chemical applications of organometallic compounds.

Emphasis in recent times has returned to new methods of organic synthesis via organometallic starting materials and intermediates. Research in this field has received a major stimulus from the discovery of several important industrial processes involving organometallic reagents and catalysts.

We believe that it is now timely to present a concise account of the applications of organometallic starting materials, intermediates, and catalysts in organic synthesis. These applications cover a wide range in terms of usefulness and generality, and in the nature of the particular atoms being linked together in the synthetic processes.

In Part I we survey the different roles that metal atoms can play in organometallic reactions and introduce a classification of organometallic syntheses based on likely or established mechanisms and metal atom functionality. We have used this classification as a means for bringing together syntheses which are closely related in terms of mechanism but which nevertheless involve quite different metal atoms.

Rather than deal with the various applications in terms of the metal involved, we have chosen to emphasize the type of bond being formed in the synthesis and have subdivided Parts II and III in this manner rather than in terms of groups or sub-groups of the Periodic Table. Thus Part II (Chapters 3–5) deals with the formation of carbon–carbon bonds and Part III (Chapters 6–7) with the formation of bonds linking carbon to other atoms, especially hydrogen, nitrogen, oxygen and other electronegative elements. Less attention is given to the synthesis of organometallic compounds themselves. Throughout the book we are concerned with reaction products which, in contrast to the intermediates or starting materials, are not organometallic in nature.

We wish to thank our colleagues, Professor W. R. Jackson, and Dr. P. W. Ford, for reading the manuscript and for their helpful comments.

Monash University,                                                    J. M. Swan
Victoria, Australia                                                   D. St.C. Black

# Contents

# Part I Some aspects of organometallic chemistry

# General introduction 1

## 1.1 Historical background

Organometallic compounds have a long history. Frankland discovered the methyl and ethyl derivatives of zinc, mercury and tin in 1849, and these compounds played an important role in the development of valence theory. They continue to be of central importance in the formulation of theories of chemical bonding and structure. Ehrlich, the chief pioneer of chemotherapy, did much research into the organic chemistry of mercury, and organomercurials have been widely used as antifungal, antibacterial and other pharmaceutical agents. Grignard's work on the organomagnesium compounds provided chemistry with a supremely powerful range of synthetic methods, being matched only now by developments in the organic chemistry of the alkali metals, boron, aluminium, nickel, copper, palladium, and many other metals. Nature also employs organometallics – reactions of the $B_{12}$ coenzymes in living matter involve the formation and cleavage of carbon–cobalt bonds.

In the broadest sense, 'metal–organic' compounds can be taken to include not only the true organometallics in which an alkyl, aryl, or cyclopentadienyl or related group is bonded to the metal by one or more metal–carbon bonds, but also the metal alkoxides and aryloxides, the metallic salts of acids, and the very large and diverse families of co-ordination compounds. All these substances are of interest to both organic and inorganic chemists, and many have special relevance to chemical technology. A great deal of current research is directed towards the development of new and useful chemical reactions through the agency of metal–organic compounds and intermediates.

In this book, however, we propose to limit the term 'organometallic' to substances having carbon-to-metal bonds. While this excludes all those metal-organic compounds where the organic group is linked to the metal via an oxygen, nitrogen, sulphur or other electronegative atom, it still leaves a large and important group of substances and reactions to be discussed. Emphasis will be placed on synthetic uses of organometallic compounds, including the role that organometallic intermediates play in many reactions. In general, where a metal or metal salt forms part of a reaction mixture, there is the possibility that an organometallic intermediate is involved in the reaction.

Sodium or lithium alkyls are intermediates in the Claisen, Michael and related reactions, and in syntheses of the acetoacetic ester and malonic ester type.

$$2CH_3C\overset{\displaystyle O}{\underset{\displaystyle OEt}{<}} \quad \xrightarrow[\text{(ii) } H_3O^\oplus]{\text{(i) NaOEt}} \quad CH_3\underset{\underset{\displaystyle O}{\|}}{C}CH_2C\overset{\displaystyle O}{\underset{\displaystyle OEt}{<}}$$

Organomagnesium compounds ('Grignard reagents') have already been mentioned as having a wide range of applications, especially the synthesis of alcohols, aldehydes, ketones, carboxylic acids, esters, and amides. Specific use has also been made of the less reactive organocadmium compounds in the synthesis of ketones. Organocobalt compounds are intermediates in the important 'oxo' process used industrially for converting alkenes into aldehydes by reaction with carbon monoxide and hydrogen in the presence of a cobalt catalyst.

$$RCH=CH_2 \quad \xrightarrow[\text{Cobalt catalyst}]{H_2/CO} \quad RCH_2CH_2CHO$$

The hydration of alkynes catalysed by mercury(II) salts in the presence of aqueous acid is thought to involve both $\pi$-bonded and $\sigma$-bonded organomercury products. The oxidative coupling of terminal alkynes to di-ynes with copper(II) acetate in pyridine proceeds via intermediate formation of the copper(I) acetylide. The Clemmensen reduction of carbonyl to methylene groups using zinc amalgam in hydrochloric acid may well proceed via intermediates held to the surface of the amalgam by carbon–zinc bonds.

$$\underset{\underset{\displaystyle O}{\|}}{\overset{\displaystyle R \quad R'}{\underset{\displaystyle C}{\diagdown\diagup}}} \quad \xrightarrow[\text{HCl}]{\text{Zn/Hg}} \quad RCH_2R'$$

The condensation of carbonyl compounds with $\alpha$-halogeno esters in the presence of zinc to give $\alpha$-hydroxy esters (Reformatsky reaction) was long thought to involve an organozinc compound; the true intermediate is now known to be a zinc enolate, with the zinc atom co-ordinated to the ester oxygen atom of the initial halogenoester.

$$\text{PhCHO} + \text{BrCH}_2\text{CO}_2\text{Et} \xrightarrow[\text{(ii) } H_3O^{\oplus}]{\text{(i) Zn}} \underset{\overset{|}{\text{OH}}}{\text{PhCHCH}_2\text{CO}_2\text{Et}}$$

An important stimulus to organometallic research was the discovery of ferrocene by Kealy and Pauson in 1951. In this compound an iron(II) atom is held between two cyclopentadienyl ($C_5H_5$) rings by chemical bonding involving the $\pi$-electrons and unoccupied $d$-orbitals in the metal. Similar 'sandwich' compounds have now been described for over thirty transition metals, including most of the rare earths, and $\pi$-bonded $C_5H_5$ is now a common ligand. A further stimulus came from the work of Brown, Wittig and others on metal hydrides, especially hydrides of lithium, boron and aluminium. These highly reactive compounds have provided new synthetic routes to organometallics, especially from easily-available alkenes, while the organometallic compounds so obtained have in turn proved to be versatile intermediates in synthesis.

The work of Reppe on the chemistry and applications of nickel tetracarbonyl has led to many novel and useful discoveries in synthesis, especially the carbonylation of alkynes and the formation of cyclooctatetraene from acetylene.

$$\text{RC} \equiv \text{CH} \xrightarrow[\text{Ni(CO)}_4]{\text{CO/H}_3O^{\oplus}} \text{RCH} = \text{CHCO}_2\text{H}$$

Finally, the investigations of Ziegler into the organic compounds of lithium and of aluminium culminated in the early 1950s in new processes for the polymerization of ethene and other alkenes at low temperatures and pressures, in cheap methods for the synthesis of long-chain primary alcohols, which are important raw materials in the detergent industry, and in new methods for the preparation of many other organometallic compounds. Triethylaluminium has become a large scale industrial chemical.

## 1.2 Modern developments

Modern developments in synthesis based on organometallic compounds or intermediates include the formation of alkanes, alcohols, amines and many

5

other products from alkylboron compounds; homogeneous hydrogenations catalysed by soluble rhodium complexes; and the trimerization of alkynes and various dimerizations of alkenes and alkadienes promoted by a wide variety of transition metal complexes. Palladium salts have achieved considerable prominence, especially in industry, for the synthesis of acetaldehyde and of vinyl acetate from ethene, and for the isomerization and oxidative dimerization of alkenes. Thallium salts are becoming widely used in synthesis by virtue of the easy formation and high reactivity of organothallium compounds. Highly reactive organocopper intermediates can be generated by the addition of copper(I) salts to organolithium or organomagnesium compounds, and these copper complexes usually show greater specificity in reaction than do the lithium or magnesium compounds from which they were derived.

Many of these new synthetic procedures have been discovered in the course of general research into the nature and properties of organometallic compounds. However, with an increased understanding of organometallic structure and reactivity has come an appreciation of the likely mechanisms of organo-metallic synthetic procedures, involving metal-free entities such as carbanions, carbonium ions, free radicals and carbenes. Some quite new bond-forming processes have also emerged, which are unique to organometallic intermediates. Rational design of new synthetic methods, based on an understanding of the role played by the metal atom in different types of organometallic compound, is now possible.

## 1.3    Outline of remaining chapters

The remaining chapter of Part I deals with the importance and versatility of organometallic compounds in synthesis, and the different roles that metal atoms can play in organometallic reactions. A classification of reaction types based on mechanism is introduced. Parts II and III deal systematically with organometallic processes for the construction of different classes of chemical bonds. Such bond formation generally involves the carbon atom attached directly to the metal atom of the organometallic starting material, and the reactions to be discussed are grouped according to the type of bond being formed, such as carbon–carbon, carbon–hydrogen, carbon–nitrogen, and so on.

Part II deals with the construction and modification of carbon–carbon bonds, including double and triple bonds, by such diverse processes as the alkylation and acylation of carbanions (Chapter 3), the isomerization and oligomerization of alkenes and alkynes (Chapter 4), and by carbon monoxide insertion (group migration) reactions (Chapter 5).

Part III deals with organometallic methods for joining carbon atoms to elements other than carbon, such as hydrogen (Chapter 6), or nitrogen, phosphorus, oxygen, sulphur, and halogen atoms (Chapter 7). Throughout, emphasis is placed on bond-making and bond breaking, and the role of the metal atom in these processes.

The very large and important class of transmetallation reactions, in which a carbon–metal bond is replaced by a bond between carbon and a different metal atom, is not dealt with in a separate chapter in this book; certain specific cases are mentioned as they arise. The major theme is the synthesis of organic molecules and the preparations and properties of organometallic compounds as such are not emphasized. These details are readily available from a variety of other books and reviews. However, it is worth mentioning that organometallic reagents are often highly reactive, invariably toxic, and should be handled with care.

# Metal atom functionality in organometallic reactions

# 2

## 2.1 Introduction

The central problem in constructing new organic molecules, or in synthesizing compounds of known structure by laboratory methods, can be summarized in the phrase 'the making and breaking of chemical bonds'. In any scheme of synthesis, existing chemical bonds in starting materials or intermediates will need to be broken, and new bonds formed, leading eventually to the desired molecular structure. For a synthesis to be both feasible and efficient the chemist needs to be aware of the nature of the bond making and bond breaking processes involved. It is our purpose in this book to concentrate on the important role that carbon–metal bonds can and do play in a wide range of synthetic procedures.

The carbon–carbon bond is clearly of first importance in organic chemistry, and the formation of such bonds will usually be required for the construction of the molecular framework of any desired product. Superimposed on or forming part of this framework there may be one or more functional groups which can be largely responsible for the unique character and reactivity of a given molecule. The introduction of functionality, including the attainment of the desired oxidation level requires the construction of other bonds such as carbon–oxygen, carbon–nitrogen, carbon–halogen and carbon–carbon double bonds and triple bonds. These bonds may sometimes be introduced in the same reaction that gives rise to the required carbon skeleton. For the construction of molecules of increasing complexity, highly specific and selective reactions are increasingly needed.

Organometallic compounds and intermediates offer a diversity of possibilities for the linking of carbon atoms by single or multiple bonds, and also the linking of carbon to a wide range of other elements such as hydrogen, oxygen, nitrogen, phosphorus, sulphur, and the halogens. In addition, organometallic reactions include processes both of wide generality and of high specificity and selectivity, and a few illustrative examples are given below.

The most versatile reactions for constructing carbon skeletons are probably those involving organomagnesium compounds and related metal–carbanion reactions. Standard procedures are available for converting a Grignard reagent, RMgX, into a primary alcohol having one or two more carbon atoms than the group R, an almost unlimited range of secondary and tertiary alcohols, ketones, and amides, and the related carboxylic acid having one more carbon atom than R (see Section 3.4).

Many examples could be quoted of the high specificity and selectivity obtainable through organometallic reactions. Thus, butadiene can be trimerized to yield mainly the all-*trans* isomer of 1,5,9-cyclododecatriene using a triethyl-aluminium–chromium(III) chloride catalyst, whereas a diethylchloroaluminium–titanium tetrachloride catalyst yields mainly the *cis-trans-trans* isomer. [1]

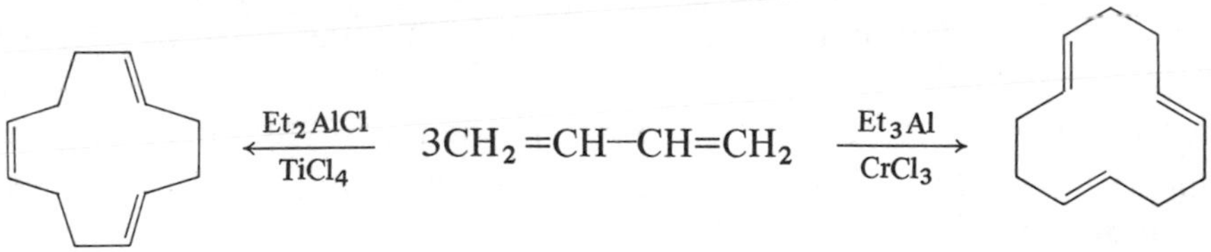

$$\xleftarrow[\text{TiCl}_4]{\text{Et}_2\text{AlCl}} \quad 3\text{CH}_2\text{=CH–CH=CH}_2 \quad \xrightarrow[\text{CrCl}_3]{\text{Et}_3\text{Al}}$$

The conversion of 1,5,9-cyclododecatriene into the all-*cis* 1,5,9-triol can be effected via an intermediate organoboron compound. This same intermediate organoborane can also yield, by reaction with carbon monoxide, a tricyclic tertiary alcohol with a new carbon skeleton [2] (see Sections 7.3.1 and 5.5).

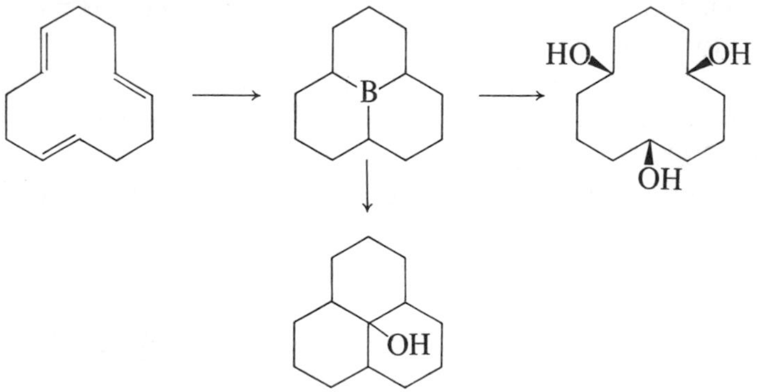

A further example of a selective process is the mercuration of nitrobenzene in the *ortho* position by mercury(II) acetate; subsequent reaction with iodine yields *o*-iodonitrobenzene. Nitrobenzene itself is resistant to direct substitution by iodine.

In general, organometallic compounds have proved to be useful reagents in organic synthesis because of the wide range of bond types and reactivities available in such compounds. Where the bond is essentially ionic, involving a metal cation associated with a carbanion, the character of the carbanion and hence its reactivity depends in part on the particular metal present. Non-ionic, or covalent carbon–metal bonds also exhibit a range of chemical reactivity, and depending on the mode of cleavage of such a bond the reactive intermediate can be either a carbonium ion, a carbanion, a free radical or a carbene-like entity.

A survey of synthetic reactions shows that the metal atom in an organometallic compound can function in at least nine different ways, as discussed below in Sections 2.2 to 2.10. The categories overlap to some extent but form a useful basis for discussion of mechanism in later chapters. In summary, the metal atom can function

a)  as a means of inducing nucleophilic (carbanionic) behaviour at a given carbon atom (Section 2.2);
b)  as a means of inducing electrophilic (carbonium ion) behaviour at a given carbon atom (Section 2.3);
c)  as a centre for the generation of a carbon free radical (Section 2.4);
d)  as a redox centre for oxidative addition and reductive elimination processes (Section 2.5);
e)  as a pivot atom which allows an intramolecular group migration from the metal to another attached atom (Section 2.6);
f)  as part of a leaving group in both substitution and elimination reactions (Section 2.7);
g)  as a centre for the generation of a carbene or carbene-like intermediate (Section 2.8);

h) as part of a metal–carbene (metallo–carbonium ion) complex (Section 2.9);
i) as a protecting or activating agent (Section 2.10).

Each of these separate functions will now be illustrated.

## 2.2    Carbanionic behaviour

The anionic part of an organometallic compound such as triphenylmethyl-
sodium, phenyllithium, or methylmagnesium iodide can function as a strong
base, i.e. as a nucleophile towards hydrogen

$$Ph_3 \overset{\ominus}{C} \overset{\oplus}{Na} + H\!-\!Y \rightarrow Ph_3 CH + \overset{\oplus}{Na} \overset{\ominus}{Y}$$

The atom Y to which the hydrogen is attached can be oxygen, nitrogen,
sulphur, phosphorus, halogen, or other electronegative element, and can also
be a carbon atom. In the latter case we observe the generation of one organometallic compound from another.

$$C_6 H_5 Li + CH_2 (CO_2 Et)_2 \rightarrow C_6 H_6 + \overset{\oplus}{Li} \overset{\ominus}{C}H(CO_2 Et)_2$$

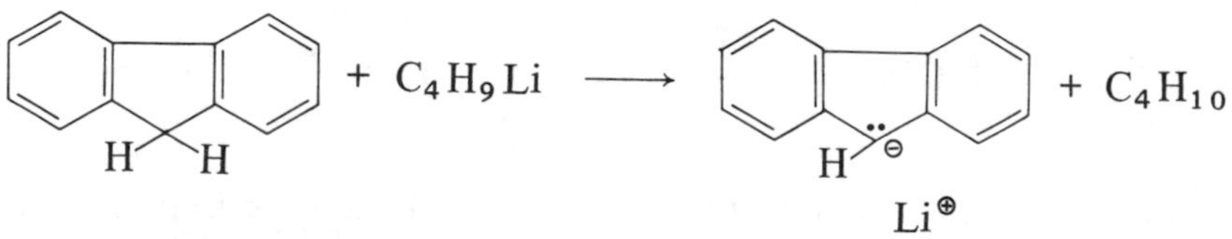

The alkali metal derivatives of enolic aldehydes, ketones, β-diketones and
β-ketoesters behave as ambident anions, with nucleophilic reactivity being
expressed at both the carbon and the oxygen atom:

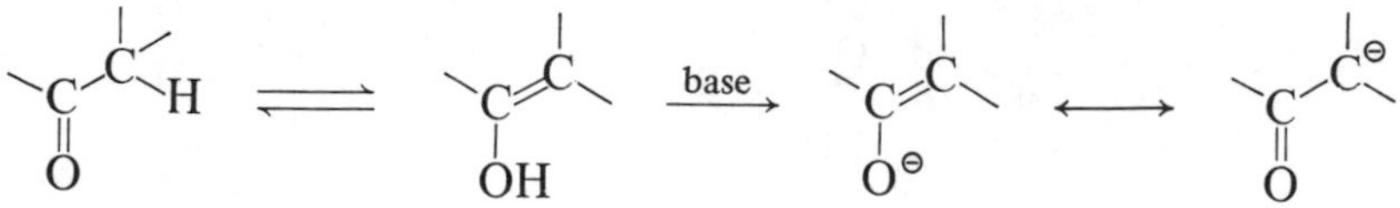

Reaction of an alkyl halide with the ambident anion derived from a β-
diketone or β-ketoester usually gives the C-alkyl derivative as major product,
but O-alkylation is also observed, as well as side reactions such as dialkylation,
oxidative coupling and Claisen condensation. [3]

11

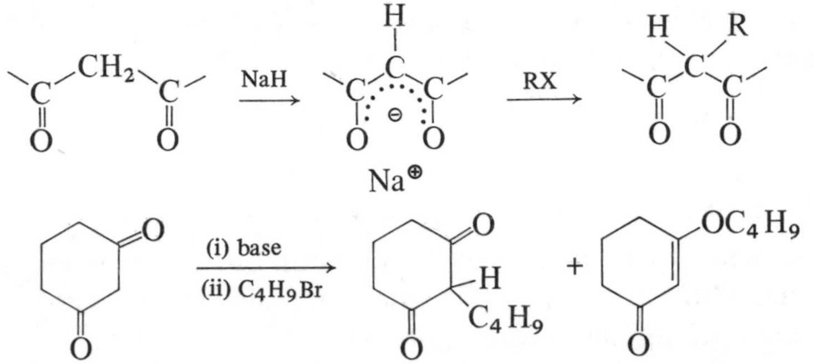

15%      37%

Carbanion-like reactivity is also found in organometallic compounds where the carbon–metal bond is essentially covalent rather than ionic. This is the case for Grignard reagents, where the covalent carbon–magnesium bond undergoes polarization and then heterolytic cleavage during reaction, but always in the sense predicted by the formula $R^{\delta\ominus}-MgX^{\delta\oplus}$, the group R behaving as a carbanion.

$$CH_3MgI + HOH \rightarrow CH_4 + IMgOH$$

$$CH_3MgI + CH_3COCH_3 \longrightarrow CH_3-\underset{\underset{OMgI}{|}}{\overset{\overset{CH_3}{|}}{C}}-CH_3 \xrightarrow{acid} (CH_3)_3COH$$

Dimethylcadmium, a liquid of b.p. 105°C, freely soluble in nonpolar organic solvents, is undoubtedly a covalent substance. Nonetheless, it reacts readily with acid chlorides to give ketones, a reaction which can be rationalized by postulating carbanion-like behaviour of the methyl groups, at least within the co-ordination complex formed between the two reactants.

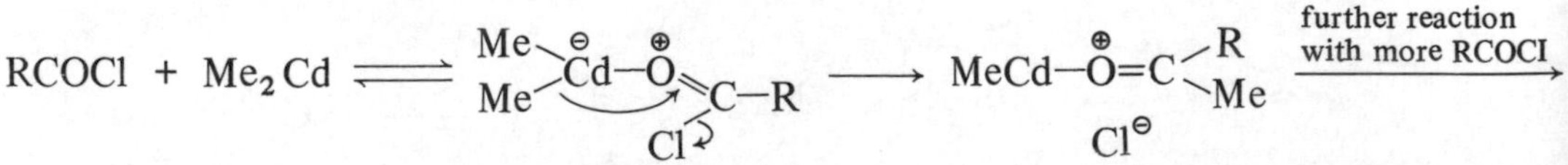

The overall reaction is

$$2RCOCl + Me_2Cd \longrightarrow 2RCOMe + CdCl_2$$

Various 'transmetallation' processes in which one organometallic compound is prepared from the halogen derivative of another metal, or from another organometallic compound, can also be rationalized in terms of carbanionic mechanisms. Some examples are:

$$2EtMgBr + CdCl_2 \longrightarrow EtCdEt + 2MgBrCl \qquad (1)$$

$$3EtMgBr + HSiCl_3 \longrightarrow HSiEt_3 + 3MgBrCl \qquad (2)$$

$$2Et_3B + 3HgCl_2 \xrightarrow[80°C]{\text{aq. NaOH}} 3Et_2Hg + (2BCl_3) \longrightarrow B(OH)_3 \quad (3)$$

$$4C_6H_5Li + Sn(CH{=}CH_2)_4 \rightleftharpoons (C_6H_5)_4Sn + 4(CH_2{=}CH)Li \quad (4)$$

For reactions of type 1 and 2 it is necessary that the metal atom in the organometallic compound be higher in the electromotive series than the metal in the metal halide. Exchange reactions of type 4 are not common, and are useful in synthesis only if the equilibrium can be displaced in the desired direction.

Further examples of carbanionic behaviour of organometallic compounds are discussed in later chapters, especially in Chapters 3 and 7.

## 2.3 Carbonium ion behaviour*

Reactions such as the hydration or bromination of alkenes are known to proceed by electrophilic attack on the electron-rich double bond to give a carbonium or bromonium intermediate which then reacts with a solvent molecule or other nucleophilic species:

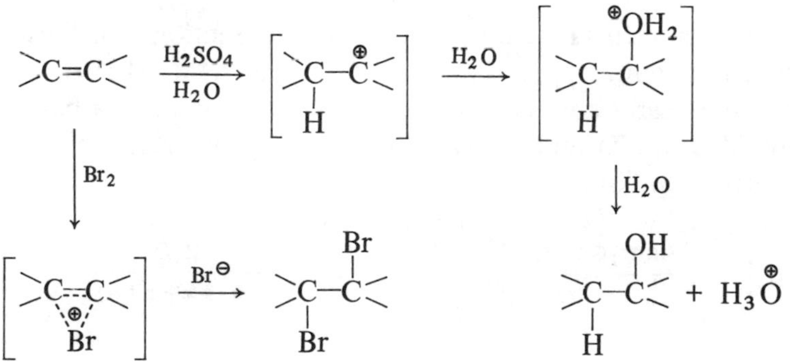

Similarly, in the methoxymercuration of alkenes a mercury(II) species attacks the double bond, presumably to yield an organometallic $\pi$-complex,

---

* The alternative term 'carbenium ion' is gaining favour for description of the species $R_3C^{\oplus}$.

which behaves as a carbonium ion in the subsequent reaction with methanol:
[4–6]

$$\text{>C=C<} \xrightarrow[\text{MeOH}]{\text{Hg(OAc)}_2} \left[ \text{>C}\cdots\text{C<} \atop \underset{\overset{|}{\text{OAc}}}{\overset{\oplus}{\text{Hg}}} \right]_{\text{AcO}^\ominus} \xrightarrow{\text{MeOH}} \overset{\text{OMe}}{\underset{\text{HgOAc}}{\text{>C–C<}}} + \text{HOAc}$$

The oxythallation of alkenes and alkynes by thallium(III) nitrate (see Section 7.5.6) can also be explained in terms of carbonium ion chemistry.

In the industrially important reaction of ethene with aqueous palladium chloride a $\pi$-bonded ethene molecule exhibits carbonium ion reactivity and undergoes attack by a hydroxyl ion, with concomitant migration of a hydride ion from one carbon to the other. The products are protonated acetaldehyde and palladium metal; in the presence of oxygen and copper(II) ions the process can be made continuous and only catalytic amounts of palladium and copper(II) salts are needed. [7] The essential step can be shown as follows.

$$\longrightarrow \quad + Pd + 2Cl^\ominus + HO^\ominus$$

$$CH_3CHO + H^\oplus$$

Further details of this reaction are discussed in Section 7.6. This area of chemistry is undergoing rapid development at present. [8–12] A recent example involving carbon–carbon bond synthesis is the reaction of a carbanion with an alkene $\pi$-bonded to platinum or palladium. The alkene again behaves as an incipient carbonium ion.

$$\overset{\text{C}}{\underset{\text{C}}{\parallel}} \longrightarrow Pd< \quad \xrightarrow{\overset{\ominus}{\text{CH(CO}_2\text{Et)}_2}} \quad \overset{\text{CH(CO}_2\text{Et)}_2}{\underset{\text{Pd<}}{\text{C}}} \quad \xrightarrow[\text{(cf. Section 6·4)}]{\text{H}_2/\text{Ni}}$$

$$\overset{\text{CH(CO}_2\text{Et)}_2}{\underset{\text{C}\diagdown\text{H}}{\text{C}}}$$

14

A so-called carbene complex of tungsten carbonyl, $Ph(MeO)C-W(CO)_5$ has been found to behave as a stabilized carbonium ion and affords a vinyl ether on undergoing reaction with a phosphorus ylid (a stabilized carbanion, see Section 3.1). [13]

$$C_6H_5 \overset{\oplus}{\underset{MeO}{}}\overset{\ominus}{C}-W(CO)_5 \ + \ \overset{R}{\underset{H}{}}\overset{\ominus}{C}-\overset{\oplus}{P} \overset{C_6H_5}{\underset{C_6H_5}{\diagdown}} C_6H_5 \ \longrightarrow$$

$$C_6H_5 \overset{}{\underset{MeO}{}}C = C \overset{-R}{\underset{H}{}} \ + \ C_6H_5 \overset{\oplus}{\underset{C_6H_5}{}}\overset{\ominus}{P}-W(CO)_5$$

cis and trans

Transition-metal carbene complexes are dealt with more fully in Section 2.9 below. In many cases it is a matter of convenience whether the organic entity attached to the metal atom in 'carbene complexes' is regarded as a stabilized carbene, $L_nM-C\overset{X}{\underset{Y}{}}$ , or as a metallo–carbonium ion, $L_n\overset{\ominus}{M}-\overset{\oplus}{C}\overset{X}{\underset{Y}{}}$ . ($L_nM$ represents the appropriate number $n$ of ligands L attached to a metal atom M.) In our view the carbonium ion formulation is more realistic and useful, despite current popularity of the term 'metal-carbene complex'.

The reaction of ethyl diazoacetate with an alkene to give a cyclopropane derivative is catalysed by various transition metals and the essential intermediate is thought to be a metal stabilized carbene. [14] However, reaction of this intermediate with the alkene can be better understood in terms of carbonium ion chemistry. For the reaction of ethyl diazoacetate with cyclohexene catalysed by an alkylphosphite–copper(I) chloride complex, the following equations can be written:

$$(RO)_3PCuCl \ + \ N_2CHCO_2Et \ \longrightarrow$$

$$N_2 \ + \ (RO)_3 P-\overset{Cl}{\underset{}{\overset{|}{Cu}}}-CHCO_2Et \ \longleftrightarrow \ (RO)_3P-\overset{Cl}{\underset{}{\overset{|}{\underset{\ominus}{Cu}}}}-\overset{\oplus}{C}HCO_2Et$$

15

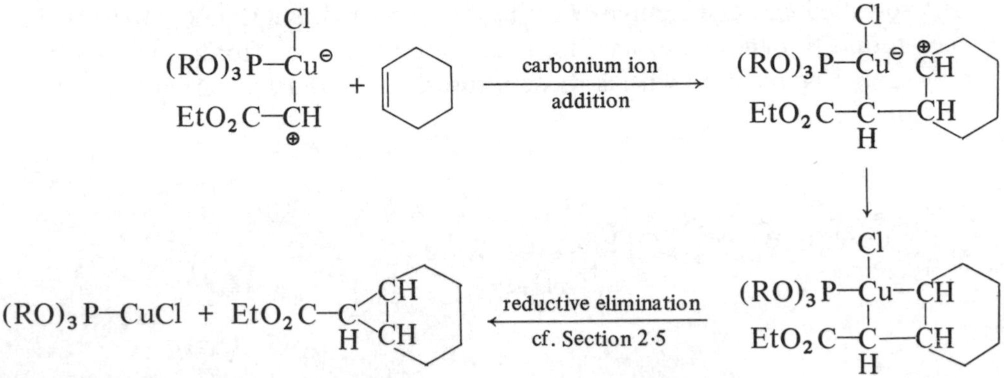

The reaction of diazomethane with benzene in the presence of copper(I) bromide to yield tropane is thought to involve a copper-stabilized methylene carbonium ion. [15]

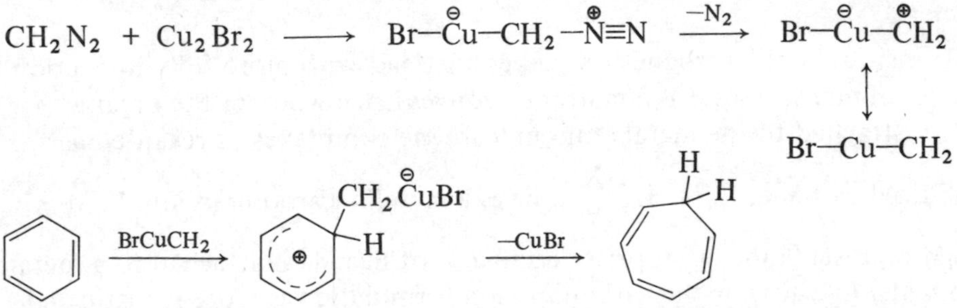

Other examples involving enhancement of electrophilic reactivity at a carbon atom resulting from metal co-ordination are mentioned in Section 2.10 below.

## 2.4 Free radical intermediates

A general method for the formation of a free radical is the one-electron oxidation of the corresponding anion. For example, oxidation of a phenolate anion can yield a phenoxy radical.

$$Ar{-}\ddot{\underset{\cdot\cdot}{O}}{:}^{\ominus} \longrightarrow Ar{-}\ddot{\underset{\cdot\cdot}{O}}{\cdot} + e^{\ominus}$$

$$\text{oxidant} + e^{\ominus} \longrightarrow \text{reductant}$$

If a carbon species, formally a carbanion, is attached to a metal atom which can readily undergo a one-electron reduction, the organic product resulting from such an electron transfer will be a free radical.

16

e.g.,

$$RCu \longrightarrow R\cdot + Cu$$

or

$$R-Cu-R \longrightarrow 2R\cdot + Cu$$

or

$$R-Cu + Cu^{2\oplus} \longrightarrow R\cdot + 2Cu^{\oplus}$$

From the viewpoint of the organic moiety initially attached to the metal, the reaction is

$$R\colon^{\ominus} \longrightarrow R\cdot + e^{\ominus}$$

The coupling of alkynes by copper(II) acetate in pyridine is thought to involve a reaction of the last type. Traces of copper(I) ion are required, and a copper(I) acetylide is probably formed first, the pyridine acting as a base. This acetylide can then undergo oxidation by copper(II) ion, the last step being coupling of the two free radicals. [16]

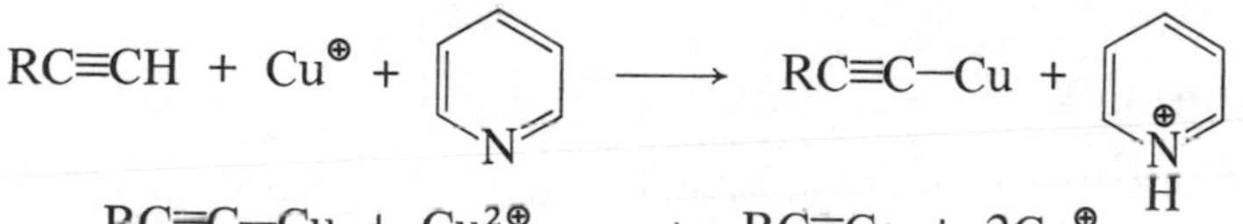

$$RC\equiv C-Cu + Cu^{2\oplus} \longrightarrow RC\equiv C\cdot + 2Cu^{\oplus}$$
$$2RC\equiv C\cdot \longrightarrow RC\equiv C-C\equiv CR$$

Aromatic Grignard reagents undergo a self-coupling reaction when treated with the halides of nickel, iron, cobalt, manganese or chromium, which act as appropriate oxidants. For the formation of diphenyl from phenylmagnesium bromide and chromium(II) chloride the following mechanism has been proposed: [17]

$$2C_6H_5MgBr + CrCl_2 \longrightarrow C_6H_5-Cr-C_6H_5 + 2MgClBr$$

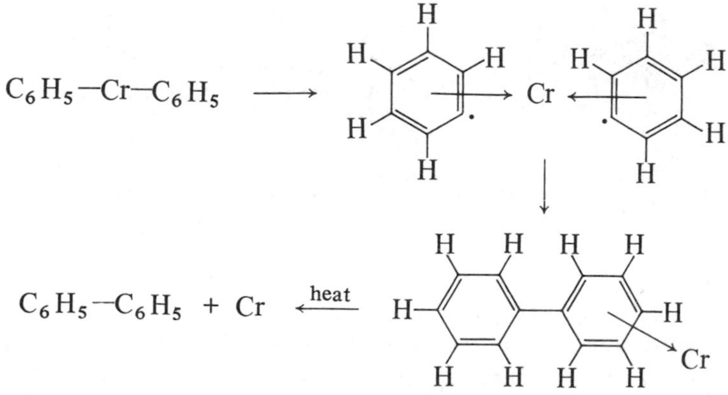

17

Reactive alkyl halides normally react with Grignard reagents to give substitution products by nucleophilic displacement of halide ion.

$$>\!C\!=\!CH\!-\!CH_2Cl + CH_3MgCl \longrightarrow >\!C\!=\!CH\!-\!CH_2CH_3 + MgCl_2$$

However, other products may be formed simultaneously by free radical processes, especially in the presence of metallic impurities. These alternative reactions are strongly favoured by the addition of catalytic amounts of cobalt(II) or copper(I) chloride. [18]

$$>\!C\!=\!CH\!-\!CH_2Cl \xrightarrow[\text{CoCl}_2]{\text{CH}_3\text{MgCl}} >\!C\!=\!CH\!-\!CH_2\!-\!CH_2\!-\!CH\!=\!C\!<$$

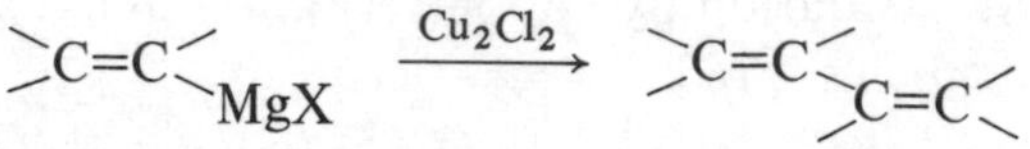

$$>\!C\!=\!C\!<_{MgX} \xrightarrow{\text{Cu}_2\text{Cl}_2} >\!C\!=\!C\!<_{C=C<}$$

$$ArMgX \xrightarrow{\text{Cu}_2\text{X}_2} ArCu \xrightarrow{\text{heat or O}_2} Ar\!-\!Ar$$

It is probable that these oxidative dimers are formed via alkyl cobalt or copper compounds, which undergo internal one-electron oxidation to give alkyl radicals. Aryl free radicals, formed from intermediate aryl copper compounds, may be involved in the Ullmann reaction of aryl halides with metallic copper. [15]

$$2C_6H_5Cl + 2Cu \longrightarrow C_6H_5\!-\!C_6H_5 + Cu_2Cl_2$$

Cross coupling reactions between aromatic Grignard reagents and vinyl halides, catalysed by various nickel(II) salts or complexes have also been reported. [19, 20]

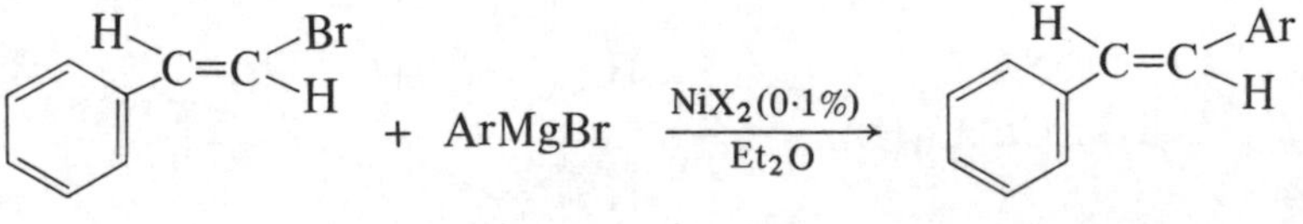

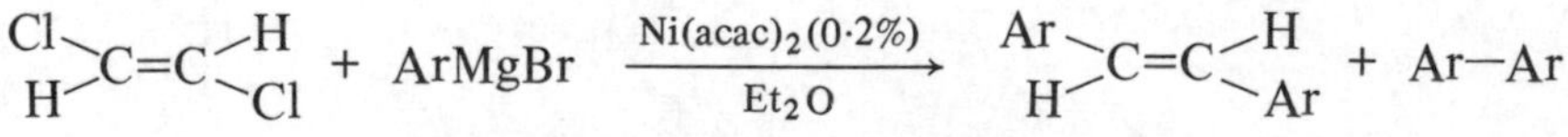

An important new class of organometallic reactions, recognized only since 1966, consists in the attack of a free radical at the metal atom of an organometallic compound, with displacement of another radical. [21] In the following equation and elsewhere $L_nMR$ represents a complex with $n$ ligands L and one ligand R attached to a metal atom M; the centre of attention is the M—R bond by which the metal atom is joined to carbon.

$$X\cdot + L_n MR \longrightarrow L_n MX + R\cdot$$

This subject has recently been reviewed. Important examples of bimolecular homolytic substitution at a metal centre are found in the reaction of alkyl derivatives of Li, Mg, Zn, Cd, B, Al, and Tl with oxygen to give the corresponding alkylperoxymetallic compounds as primary product. These reactions were initially thought to involve nucleophilic attack of $O_2$ at the metal and a 1,3-nucleophilic shift of the alkyl group from metal to oxygen:

$$L_n MR + O_2 \longrightarrow L_n \overset{\ominus}{M} \underset{O-O}{\overset{R}{\diagdown}}_{\oplus} \longrightarrow L_n M-O-O-R$$

However, it is now established that these are free radical chain reactions, and are best formulated as follows:

*Initiation:* $\qquad X\cdot + L_n MR \longrightarrow L_n MX + R\cdot$
$\qquad\qquad\qquad$ (initiator)

*Propagation:* $\qquad R\cdot + O_2 \longrightarrow R-O-O\cdot$
$\qquad\qquad ROO\cdot + L_n MR \longrightarrow ROOML_n + R\cdot$

*Termination:* $\qquad 2ROO\cdot \longrightarrow$ non-radical products

Specific examples involving compounds of aluminium, magnesium and zinc are given in Section 7.2.3.

The various metal exchange reactions of type

$$L_n M + M' \longrightarrow L_n M' + M$$

in which the metal in the organometallic compound must be lower in the electromotive series than the metal initiating the exchange, should probably be included in this section. Thus metallic sodium, with one valence electron, can be regarded as analogous to a free radical, and the exchange reaction

$$Et_2 Hg + 2Na \longrightarrow 2EtNa + Hg$$

can be written

$$Et-Hg-Et + Na \longrightarrow EtHgNa + Et\cdot$$
$$EtHgNa + Na \longrightarrow Et\cdot + Na_2 Hg$$
$$2Et\cdot + 2Na \longrightarrow 2EtNa$$

_____

Sum: $\qquad Et_2 Hg + 4Na \longrightarrow 2EtNa + Na_2 Hg$

Organoboranes react readily with acrolein, methyl vinyl ketone, and similar compounds, and these reactions are likewise free radical in nature. Hydrolysis of the borate enol ester intermediate yields a saturated aldehyde as product.

$$X \cdot (\text{initiator}) + R_3 B \longrightarrow R_2 BX + R \cdot$$

$$R \cdot + CH_2 = CH - CHO \longrightarrow RCH_2 - \overset{\cdot}{C}H - CHO \longleftrightarrow RCH_2 - CH = CH - O \cdot$$

$$RCH_2 - CH = CH - O \cdot + R_3 B \longrightarrow RCH_2 - CH = CH - OBR_2 + R \cdot$$

$$RCH_2 - CH = CH - OBR_2 \xrightarrow{H_2O} RCH_2 - CH = CH - OH \rightleftharpoons RCH_2 CH_2 CHO$$

A further reaction which may involve free radicals is the oxidative coupling of allyl acetates with tetracarbonylnickel. [22]

$$2 RCH = CHCH_2 OCOCH_3 \xrightarrow{Ni(CO)_4} RCH = CHCH_2 CH_2 CH = CHR$$

Radical anions are also implicated in a number of organometallic processes, especially those involving metallic sodium (see Sections 3.2 and 4.3.2). The reductions of ketones and of various alkyl and aryl halides with organometallic hydrides are also thought to be free radical in character (see Section 6.6). Aryl radicals can be generated by the photolysis of $ArTl(OCOCF_3)_2$ by oxidation of this or related thallium(III) compounds with palladium chloride, or by reaction of aryl magnesium halides with thallium(I) bromide. [23–25]


## 2.5    Oxidative addition and reductive elimination processes

A new general method for the synthesis of carbon–carbon bonds under mild, neutral reaction conditions is based upon the 'oxidative addition' of organic species to a co-ordinatively unsaturated diamagnetic transition metal compound to yield a new complex in which two organic groups are joined to the metal atom by $\sigma$-bonds. 'Reductive elimination' from this complex, e.g., by the action of heat, yields a product with a new carbon–carbon bond. The initial organotransition metal complex can be shown as $L_n M^{m-2} - R$, in which the formal oxidation number of the metal atom is $m - 2$, the metal atom possessing a total of 16 valence electrons. Oxidative addition of $R'X$ yields a new complex in which the formal oxidation number of the metal atom is now $m$, (18 valence electrons). Thus,

$$L_n M^{m-2} - R + R'X \longrightarrow L_n M^m \overset{\displaystyle R}{\underset{\displaystyle X}{\diagdown} R'}$$

Reductive elimination of the groups R and R′ results in the appearance of the new product R—R′, with return of the oxidation number of the metal atom to the initial $m - 2$ state. [26]

A specific example is shown below.

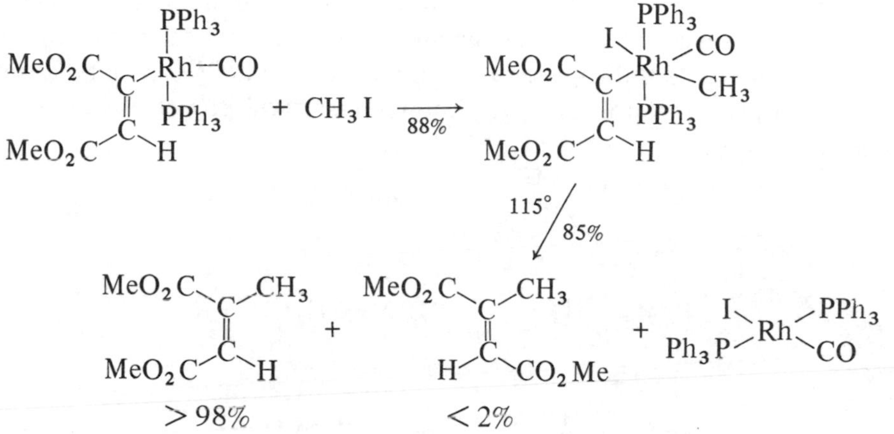

Further examples are found in Sections 2.9, 5.4.2 and 7.7.

In the above example it is thought that the carbon–carbon coupling step is concerted, and that free radicals are not involved. The initial complex is prepared as follows:

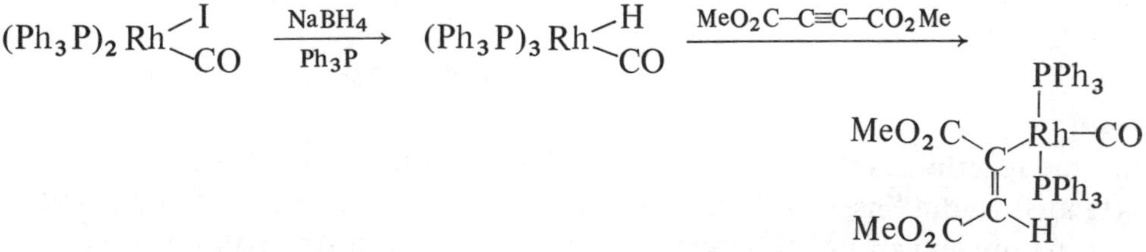

The process of reductive elimination from a co-ordinated metal atom can also be applied to the synthesis of carbon–halogen and carbon–hydrogen bonds.

$$L_nM\begin{smallmatrix}R\\X\end{smallmatrix} \longrightarrow L_nM + RX$$

$$L_nM\begin{smallmatrix}R\\H\end{smallmatrix} \longrightarrow L_nM + RH$$

Specific examples are given in Sections 2.9, 5.2, 5.4.2, 6.3., and 7.8.

## 2.6    Group migration (insertion) reactions

The so-called insertion reactions of metal complexes form a clearly defined and important mechanistic class. In these reactions an entity such as carbon monoxide, an alkene, or an alkyne co-ordinates to a metal atom and then appears to insert itself into an existing metal–carbon bond of that complex. Thus if $L_n$ represents the appropriate number of ligands, M is a metal atom and R is an organic group, some well known 'insertion' processes can be shown as follows:

$$L_n M \overset{CO}{\underset{R}{\big<}} \longrightarrow L_n M - \underset{\underset{O}{\|}}{C} - R$$

$$L_n M \overset{\overset{CH_2}{\|}{CH_2}}{\underset{R}{\big<}} \longrightarrow L_n M - CH_2 CH_2 - R$$

$$L_n M \overset{\overset{CH}{\|}{CH}}{\underset{R}{\big<}} \longrightarrow L_n M - CH = CH - R$$

These processes actually represent a migration of the group R from the metal atom to a carbon atom of the co-ordinated carbon monoxide, alkene, or alkyne. [27] In the case of carbonyl insertion this can be written

$$-\underset{|}{M} \overset{C \overset{O}{\|}}{\underset{R}{\big<}} \longrightarrow -\underset{|}{M} - \overset{\overset{O}{\|}}{C} \diagdown R$$

and the reaction can be described as a 1,2-migration. This definition can be extended to the 'insertion' reactions of sulphur dioxide, on the assumption that the organic group, R, migrates from a metal atom to a co-ordinated sulphur atom. [28] However these reactions may be free radical in character (cf. Section 2.4). [29]

Thus (insertion): $\qquad L_n M - R \xrightarrow{SO_2} L_n M \overset{S \overset{O}{\diagup} \diagdown O}{\underset{R}{\big<}} \longrightarrow L_n M - \overset{\overset{O}{\|}}{\underset{\underset{R}{|}}{S}} = O$

or (free radical)

Initiation: $\qquad L_n MR + X \cdot \longrightarrow L_n MX + R \cdot$

Propagation: $\qquad R \cdot + SO_2 \longrightarrow RSO_2 \cdot$

$\qquad RSO_2 \cdot + L_n MR \longrightarrow RSO_2 ML_n + R \cdot$

Group migration reactions leading to the formation of carbon–carbon bonds are especially important, and are exemplified below by the 'oxo' process for the industrial preparation of butanal from propene, the polymerization of ethene by triethylaluminium and the various carbonylation reactions of trialkylboranes.

The 'oxo' synthesis of butanal involves the addition of carbon monoxide and hydrogen to propene in the presence of octacarbonyldicobalt. The true catalyst is tricarbonylhydrido cobalt, $HCo(CO)_3$, which undergoes addition to the propene; the propyl group subsequently migrates to the carbon atom of a co-ordinated carbon monoxide molecule and the resulting intermediate undergoes hydrogenolysis to the product aldehyde (see also Section 5.2):

$$CH_3 CH{=}CH_2$$

$$+$$

$$HCo(CO)_3 \longrightarrow CH_3 CH_2 CH_2{-}Co(CO)_3 \xrightarrow{\ CO\ } CH_3 CH_2 CH_2{-}Co(CO)_4$$

$$HCo(CO)_3 + CH_3 CH_2 CH_2 C{\overset{H}{\underset{O}{<}}} \xleftarrow{\ H_2\ } CH_3 CH_2 CH_2{-}\overset{}{\underset{\overset{\|}{O}}{C}}{-}Co(CO)_3$$

The first step of this catalytic reaction, the addition of the cobalt hydride to the alkene double bond, can also be shown as a migration, the hydrogen atom moving (as a hydride ion) from the metal atom to the more substituted carbon of the co-ordinated alkene:

$$CH_3 CH{=}CH_2 + HCo(CO)_3 \longrightarrow$$

$$CH_3 CH{\overset{CH_2}{<}}\quad \begin{matrix} & CO \\ Co & CO \\ H & CO \end{matrix} \longrightarrow$$

$$CH_3 CH_2 {\overset{CH_2}{\diagdown}} Co {\overset{CO}{<}}{\underset{CO}{CO}}$$

The polymerization of ethene by triethylaluminium, the so called 'growth' reaction in which each ethyl group initially attached to aluminium grows to a long alkyl chain by successive additions of ethene units can be best understood as a migration (insertion) reaction (see also Section 4.4.3)

23

$$\underset{Et}{\overset{Et}{\diagdown}}Al-CH_2CH_3 + CH_2{=}CH_2 \rightleftharpoons$$

$$\underset{Et}{\overset{Et}{\diagdown}}Al\underset{CH_2CH_3}{\overset{\overset{CH_2}{\|\quad}{CH_2}}{\diagup}} \longrightarrow \underset{Et}{\overset{Et}{\diagdown}}\overset{\ominus}{Al}\underset{CH_2CH_3}{\overset{CH_2\overset{\oplus}{C}H_2}{\diagup}}$$

$$Et(CH_2CH_2)_l-Al\underset{(CH_2CH_2)_n\,Et}{\overset{(CH_2CH_2)_m\,Et}{\diagup}} \xleftarrow[\text{etc.}]{CH_2{=}CH_2} \underset{Et}{\overset{Et}{\diagdown}}Al-CH_2CH_2CH_2CH_3$$

If the dipolar structure $Et_2\overset{\ominus}{Al}\underset{CH_2CH_3}{\overset{CH_2\overset{\oplus}{C}H_2}{\diagup}}$ is valid for the presumed intermediate, the subsequent migration of an ethyl group from aluminium to
carbon is a 1,3- rather than a 1,2-migration.

In the carbonylation of trialkylboranes, $R_3B$, the carbon atom of carbon
monoxide initially becomes attached to the boron atom. Alkyl groups then
migrate intramolecularly from boron to carbon, and experimental results are
consistent with the following stepwise mechanism. [30]

$$R_3B + CO \rightleftharpoons R_3\overset{\ominus}{B}-C{\equiv}\overset{\oplus}{O} \longrightarrow \underset{\overset{\|}{O}}{R_2B-C-R}$$

$$O{=}B-\underset{R}{\overset{R}{C}}{\diagup}{\diagdown}R \longleftarrow R-B\underset{O}{\overset{}{\diagup}}C\overset{R}{\underset{R}{\diagdown}}$$

By suitable choice of reaction conditions and according to the nature of the
alkyl groups R, it is possible to control this reaction to allow the migration of
one, two, or three R groups, and the final products, after appropriate work-
up, can be variously aldehydes, ketones, or primary, secondary or tertiary
alcohols (see Section 5.5).

Another possible example of group migration is found in the reaction of
arylmercury halides or diarylmercury compounds with alkenes in the presence
24

of palladium, rhodium or ruthenium salts to yield aryl substituted alkenes. In the case of palladium chloride, diphenylmercury, and ethene in the presence of lithium chloride, the reaction involves the intermediate formation of a phenylpalladium complex which co-ordinates to ethene. Phenyl migration from palladium to carbon yields a phenylethylpalladium salt which forms styrene by elimination of a palladium hydride (see Section 2.7 below) which decomposes further to palladium metal.

$$PhHgPh + LiPdCl_3 \longrightarrow Li^{\oplus}[PhPdCl_2]^{\ominus} + PhHgCl$$

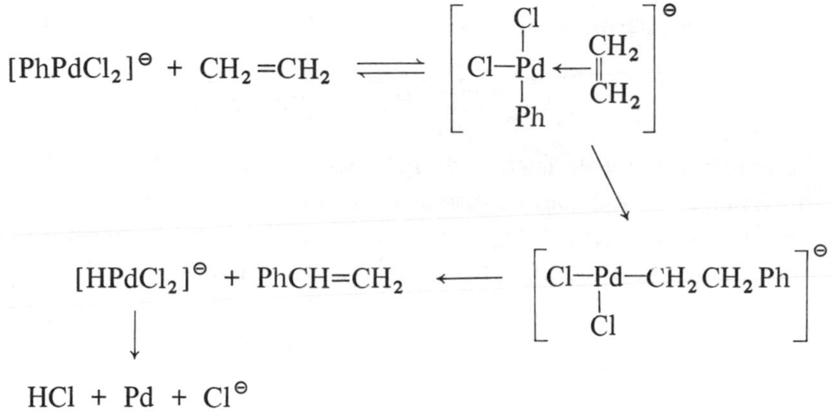

$$[PhPdCl_2]^{\ominus} + CH_2{=}CH_2 \rightleftharpoons \left[ \begin{array}{c} Cl \\ | \\ Cl{-}Pd{\leftarrow}\|{\overset{CH_2}{\underset{CH_2}{}}} \\ | \\ Ph \end{array} \right]^{\ominus}$$

$$[HPdCl_2]^{\ominus} + PhCH{=}CH_2 \longleftarrow \left[ \begin{array}{c} Cl{-}Pd{-}CH_2CH_2Ph \\ | \\ Cl \end{array} \right]^{\ominus}$$

$$HCl + Pd + Cl^{\ominus}$$

Similar reactions with a wide variety of other alkenes, including allyl alcohols, allyl halides, enol esters, ethers, and vinyl halides have been reported. [31]

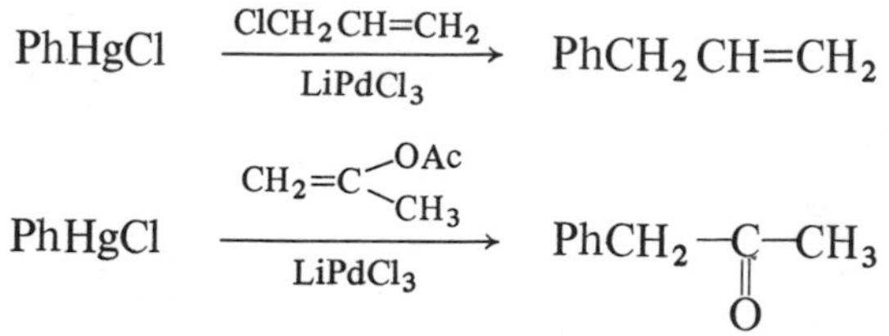

$$PhHgCl \xrightarrow[LiPdCl_3]{ClCH_2CH{=}CH_2} PhCH_2CH{=}CH_2$$

$$PhHgCl \xrightarrow[LiPdCl_3]{CH_2{=}C{\overset{OAc}{\underset{CH_3}{}}}} PhCH_2{-}\underset{\underset{O}{\|}}{C}{-}CH_3$$

## 2.7    Leaving groups in substitution and elimination reactions

Grignard reagents and related organometallic compounds react readily with dilute acids to yield hydrocarbons.

$$R-MgBr + H_3O^\oplus \longrightarrow RH + H_2O + BrMg^\oplus$$

$$C_6H_5Li + H_3O^\oplus \longrightarrow C_6H_6 + H_2O + Li^\oplus$$

$$C_6H_5HgOAc + H_3O^\oplus \longrightarrow C_6H_6 + H_2O + AcOHg^\oplus$$

These substitution reactions can be regarded as electrophilic attack of the hydrated proton on an anionic or potentially anionic carbon atom, the metal atom forming part of the leaving group.

Other electrophiles can react similarly, and the reaction of an arylmercury salt with iodine to yield an aryl iodide can be regarded as analogous to the electrophilic substitution of an aromatic hydrocarbon (an arene) by a halogen.

$$ArHgI + I_2 \longrightarrow ArI + HgI_2$$

$$ArH + X_2 \longrightarrow ArX + HX$$

The reaction between activated aryltrimethylsilanes and acid chlorides in the presence of aluminium chloride leads to the formation of ketones. This can be viewed as electrophilic attack by an acid chloride/aluminium chloride complex on the aryl group, the leaving group being the trimethylsilyl cation. [32]

$$ArSiMe_3 + RCOCl \xrightarrow{AlCl_3} ArCOR + Me_3SiCl$$

In parallel to these substitution reactions, elimination reactions are known in which the cationic leaving group contains a metal atom such as Li, Na, Mg, B, or Al [33] (see Section 4.1.1).

$$CH_3-CH_2Li \xrightarrow{heat} CH_2=CH_2 + LiH$$

The kinetics of this pyrolysis reaction are first order and a possible mechanism can be written [34]

$$\underset{H}{\overset{H}{\diagdown}}C \cdots C \underset{Li}{\overset{H}{\diagup}}H \longrightarrow \underset{H}{\overset{H}{\diagdown}}C=C\underset{H}{\overset{H}{\diagup}} + LiH$$

An elimination reaction of major industrial importance is the formation of terminal alkenes by the pyrolysis of trialkylalanes.

$$R_2AlCH_2CH_2R' \longrightarrow R_2AlH + CH_2=CHR'$$

Carbon–metal bonds can also be broken by various reductive procedures, including hydrogenolysis.

$$R_3B \xrightarrow{H_2} R_2BH + RH$$

$$RHgX \xrightarrow{NaBH_4} RH$$

In these cases, the metal atom can be regarded as a 'leaving group' but it does not necessarily follow that bond breaking is initiated by attack of a nucleophile (e.g. $H^\ominus$) at the carbon atom attached to the metal. If the reaction is free radical in character, initial attack at the metal is more probable (cf. Section 7.4.4).

$$RHgX \xrightarrow{NaBH_4} RHgH \longrightarrow R\cdot + \cdot HgH$$

$$R\cdot + HHg\cdot \longrightarrow RH + Hg$$

If the 2-hydroxy-1-acetoxymercury compound formed from a terminal alkene and aqueous mercury(II) acetate is treated with palladium chloride, the product is the corresponding methyl ketone. [35]

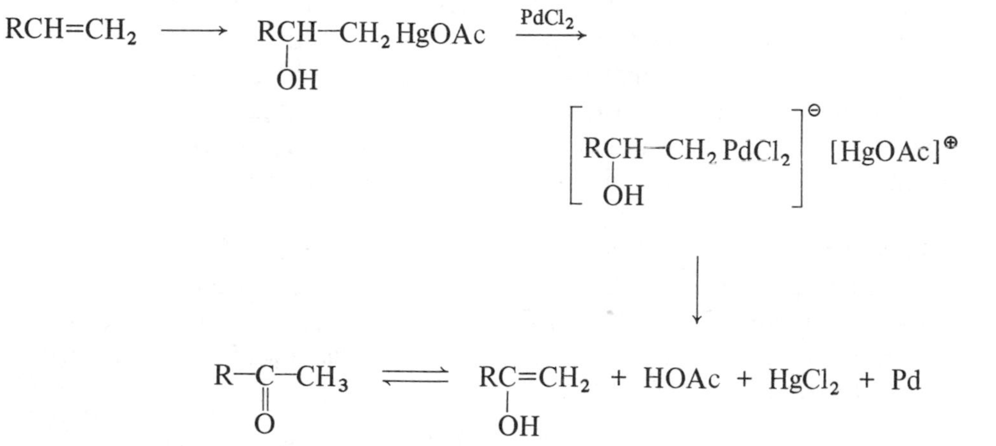

$$RCH=CH_2 \longrightarrow \underset{\underset{OH}{|}}{RCH}-CH_2HgOAc \xrightarrow{PdCl_2}$$

$$\left[ \underset{\underset{OH}{|}}{RCH}-CH_2PdCl_2 \right]^{\ominus} [HgOAc]^{\oplus}$$

$$\underset{\underset{O}{\|}}{R-C}-CH_3 \rightleftharpoons \underset{\underset{OH}{|}}{RC}=CH_2 + HOAc + HgCl_2 + Pd$$

Decomposition of the σ-bonded organopalladium complex probably involves the initial elimination of a palladium(II) hydride, which decomposes further to palladium metal.

27

$$\left[\begin{array}{c} H \\ R-\overset{|}{\underset{|}{C}}-CH_2-PdCl_2 \\ OH \end{array}\right]^{\ominus} \longrightarrow \underset{HO}{\overset{R}{>}}C=CH_2 + [HPdCl_2]^{\ominus}$$

$$R-\underset{\underset{O}{\|}}{C}-CH_3 \qquad HCl + Pd + Cl^{\ominus}$$

The action of a variety of bases on 2-hydroxy-1-chloromercury compounds has been shown to yield oxirans (epoxides). Here again the mercury functions as part of a leaving group in an internal solvolysis reaction. [36]

$$\begin{array}{ccc} & \overset{OH}{\underset{\big\downarrow}{}} & \\ H_2C & \overset{CH}{\diagdown} & \overset{\cdots HgCl}{CH} \\ H_2C & \!\!-\!\! & CH_2 \end{array} \longrightarrow \begin{array}{ccc} & \overset{H}{\underset{C-O}{}} & \\ H_2C & & CH \\ H_2C & \!\!-\!\! & CH_2 \end{array}$$

## 2.8   Carbene and carbenoid intermediates

Carbenes ($H_2C:$, $R_2C:$, $Cl_2C:$, etc.) are highly reactive two-co-ordinate carbon species having six valence electrons associated with the central carbon atom. The two non-bonding electrons may be either paired or unpaired, corresponding to the spectroscopic singlet or triplet states respectively. Practically all carbenes have a lifetime considerably less than one second, and many different methods for their generation are known. The organometallic intermediates which are of interest here can either give rise to 'free' carbenes or can act as 'carbenoid' precursors.

Phenyl(trihalogenomethyl)mercury compounds, $PhHgCX_3$, where X = Cl or Br, are reagents of outstanding utility for the preparation of 1,1-dihalogenocyclopropanes from alkenes. Free dihalogenocarbenes are formed by elimination of phenylhalogenomercury and undergo addition to the carbon–carbon double bond. [37, 38]

$$PhHgCX_3 \xrightarrow{\text{heat}} PhHgX + [:CX_2]$$

$$>\!C=C\!< + [:CX_2] \longrightarrow \underset{X\diagup \overset{|}{C}\diagdown X}{>\!C\!-\!C\!<}$$

28

Elimination of phenylbromomercury is much more favourable than that of phenylchloromercury, and consequently phenyl(bromodichloromethyl)-mercury is a valuable reagent for the generation of dichlorocarbene.

$$PhHgCCl_2Br \; + \; \text{[cyclohexene]} \quad \xrightarrow[\text{2h}]{\text{benzene, 80°C}} \quad PhHgBr \; + \; \text{[bicyclic]}\begin{array}{c} Cl \\ Cl \end{array}$$

The extrusion of dichlorocarbene from phenyl(bromodichloromethyl)-mercury is thought to be concerted, proceeding via a cyclic transition state.

$$\begin{array}{c} :Br \\ Ph-Hg-C-Cl \\ | \\ Cl \end{array} \quad \text{or} \quad \begin{array}{c} Ph-Hg \cdots Br \\ \vdots \\ C-Cl \\ | \\ Cl \end{array}$$

This picture provides a satisfactory explanation for the preferential elimination of phenylbromomercury since intramolecular attack at mercury by bromine should be more favourable than attack by chlorine, and the carbon–bromine bond is weaker than the carbon–chlorine bond.

On the other hand, the related reactions of bis(bromomethyl)mercury with alkenes do not proceed via free methylene ($:CH_2$); experimental evidence favours a mechanism involving direct reaction between the mercurial and the alkene.

$$Hg(CH_2Br)_2 \; + \; \text{[cyclohexene]} \quad \longrightarrow \quad BrCH_2HgBr \; + \; \text{[bicyclic]}\begin{array}{c} H \\ H \end{array}$$

Phenyl(trihalogenomethyl)mercury compounds can also be used for the transfer of dihalogenocarbenes to the multiple bonds of alkynes, imines, thioketones and ketones as well as to a variety of single bonds such as C—H, Si—H, Si—C, B—C, into which the dihalogenocarbene is inserted. The reaction of phenyl(trihalogenomethyl)mercury compounds with triphenylphosphine and carbonyl compounds to yield alkenes provides an alternative to the Wittig alkene synthesis. [39]

29

$$RCHO + PhHgCCl_3 + Ph_3P \longrightarrow RCH=CCl_2 + PhHgCl + Ph_3PO$$

Certain trihalogenomethyl tin compounds can also serve as a source of dihalogenocarbenes and in particular the formation of difluorocarbene by this method is of considerable value. [40, 41]

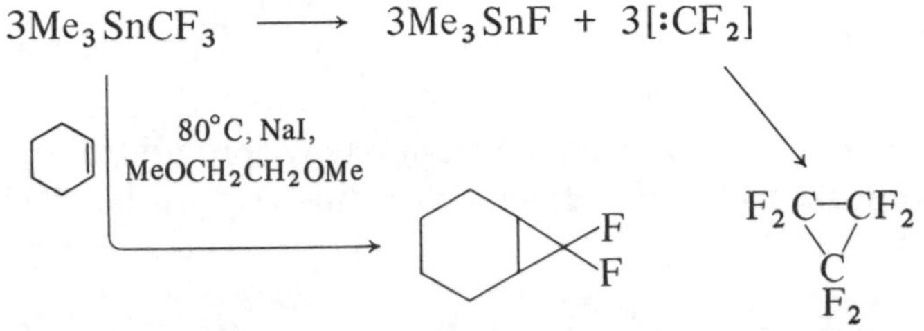

$$3Me_3SnCF_3 \longrightarrow 3Me_3SnF + 3[:CF_2]$$

The Simmons–Smith method for the preparation of cyclopropanes from alkenes involves treatment of the alkene with methylene iodide and a zinc–copper alloy. [42–44] In this case free methylene, $:CH_2$, is not involved; the attacking species is thought to be an organozinc complex, possibly $(ICH_2)_2Zn . ZnI_2$. Improved results are obtained by replacement of the zinc–copper couple with a mixture of zinc dust and a copper(I) halide. [45]

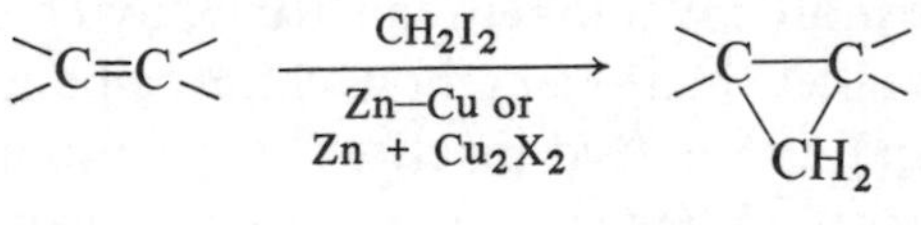

A similar carbenoid intermediate can be generated from methylene iodide and diethylzinc or by the action of diazomethane on a dialkylchloroaluminium compound. [46, 47]

$$2CH_2I_2 + 2Et_2Zn \longrightarrow (ICH_2)_2Zn–ZnEt_2 + 2EtI$$

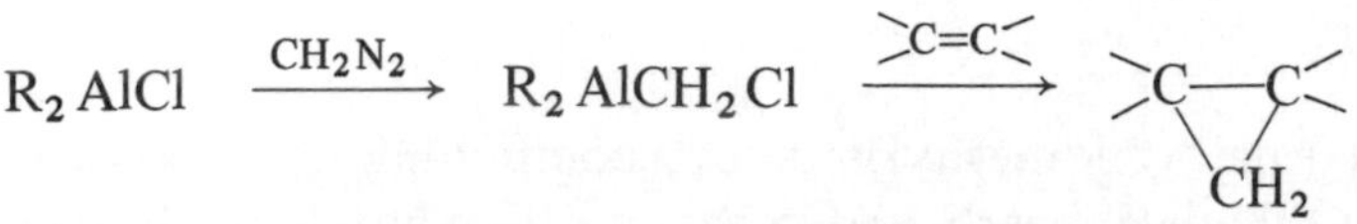

Carbenes add to a wide variety of carbon–carbon double bonds, and also to aromatic systems and to triple bonds. In so far as they have been reported to add to C=N systems, the reactions lead to carbon–nitrogen as well as carbon–carbon bond synthesis.

The so-called metal carbene complexes (see Sections 2.3 and 2.9) do not

30

generally serve as sources of free carbenes, but behave rather as metallo–carbonium ions.

## 2.9 Metal-carbene (metallo–carbonium ion) complexes as reagents and reaction intermediates

Many examples are now known of stable metal carbene complexes, of general formula $L_n M-C\underset{Y}{\overset{X}{\diagup}}$, in which a trigonal ($sp^2$ hybridized) carbon atom is attached to the metal atom and to two other univalent atoms or groups X and Y. [48] As pointed out in Sections 2.3 and 2.8 above, these complexes tend to behave as metal-stabilized carbonium ions, $L_n \overset{\ominus}{M}-\overset{\oplus}{C}\underset{Y}{\overset{X}{\diagup}}$.

These carbene complexes form a new class of organometallic substance and show a variety of interesting and useful properties. Two reactions of a typical metal–carbene giving rise to an organic product are shown below. [49, 50]

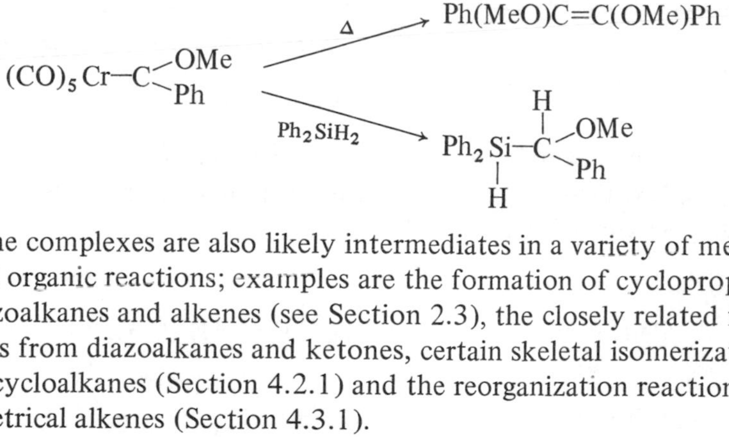

$$(CO)_5 Cr-C\underset{Ph}{\overset{OMe}{\diagup}} \quad\overset{\Delta}{\nearrow}\quad Ph(MeO)C=C(OMe)Ph$$

$$\underset{Ph_2SiH_2}{\searrow}\quad Ph_2 Si-\underset{H}{\overset{H}{\underset{|}{\overset{|}{C}}}}\underset{Ph}{\overset{OMe}{\diagup}}$$

Carbene complexes are also likely intermediates in a variety of metal-catalysed organic reactions; examples are the formation of cyclopropanes from diazoalkanes and alkenes (see Section 2.3), the closely related formation of oxirans from diazoalkanes and ketones, certain skeletal isomerizations of strained cycloalkanes (Section 4.2.1) and the reorganization reactions of unsymmetrical alkenes (Section 4.3.1).

A general form of the alkene reorganization reaction can be shown

$$X_2C=CX_2 + Y_2C=CY_2 \rightleftharpoons 2X_2C=CY_2$$

It has been suggested recently that key intermediates for such a reaction catalysed by a transition metal are the two metal–carbene complexes $L_n M-CX_2$ and $L_n M-CY_2$. [51] Each of these can react respectively with the appropriate alkene $Y_2C=CY_2$ and $X_2C=CX_2$ to yield in each case an unsymmetrical three-

31

carbon metallocyclic intermediate. Each of these intermediates can then decompose to generate the 'other' carbene complex and the unsymmetrical alkene.

$$X_2C=CX_2 \text{ and } Y_2C=CY_2 \xrightarrow{L_nM} L_nM-CX_2 \text{ and } L_nM-CY_2$$

$$L_nM-CX_2 \underset{-Y_2C=CY_2}{\overset{+Y_2C=CY_2}{\rightleftarrows}} L_nM\overset{CY_2}{\underset{CX_2}{\diagdown}}CY_2 \rightleftarrows L_nM-CY_2 + X_2C=CY_2$$

$$L_nM-CY_2 \underset{-X_2C=CX_2}{\overset{+X_2C=CX_2}{\rightleftarrows}} L_nM\overset{CX_2}{\underset{CY_2}{\diagdown}}CX_2 \rightleftarrows L_nM-CX_2 + Y_2C=CX_2$$

The formation and decomposition of a four-membered metallocyclic intermediate from the presumed metal–carbene complex, e.g.,

$$L_nM-CX_2 + Y_2C=CY_2 \rightleftarrows L_nM\overset{CX_2}{\underset{CY_2}{\diagdown}}CY_2$$

can be readily understood in terms of the 'carbonium ion' formulation of the metal–carbene (cf. Section 2.3).

$$L_nM^{\ominus}-CX_2^{\oplus} + Y_2C=CY_2 \rightleftarrows \begin{array}{c} Y_2\overset{\oplus}{C}-CY_2 \\ L_nM-CX_2 \\ \ominus \end{array} \rightleftarrows \begin{array}{c} Y_2C-CY_2 \\ | \quad\quad | \\ L_nM-CX_2 \end{array}$$

Tetracarbonylnickel and organolithium compounds combine to form unstable complexes which, however, are important synthetic intermediates, [52] (cf. also Sections 5.4.2 and 7.7).

$$RLi + Ni(CO)_4 \longrightarrow Li^{\oplus}\left[R-\overset{\overset{\textstyle O}{\parallel}}{C}-Ni(CO)_3\right]^{\ominus}$$

The acyl nickel carbonyl anion can be written with the negative charge localized on the acyl oxygen atom and in this form has been described as an 'anionic carbene complex', its reactions being discussed in relation to those of the neutral carbene complexes.

$$\left[R-\overset{\overset{\textstyle C}{\parallel}}{C}-Ni(CO)_3\right]^{\ominus} \longleftrightarrow \left[R-\overset{\overset{\textstyle O^{\ominus}}{|}}{C}-Ni(CO)_3\right]$$

However, the reactions of this and many related acylmetallates are probably best explained by ignoring the supposed relationship to a metal–carbene complex and assuming that attack by an electrophile occurs initially at the (negatively-charged) metal atom, the final organic product being formed as the result of a solvent-induced reductive elimination process from this further intermediate (cf. Section 2.5).

$$
\left[ R\overset{\overset{O}{\|}}{-}C-Ni(CO)_3 \right]^{\ominus} \xrightarrow{H^{\oplus}} R\overset{\overset{O}{\|}}{-}C-\underset{\underset{H}{|}}{Ni(CO)_3} \xrightarrow{\text{solvent}} R\overset{\overset{O}{\|}}{-}C-H + (\text{solvent})Ni(CO)_3
$$

$$
\downarrow R'Br
$$

$$
Br^{\ominus} + R\overset{\overset{O}{\|}}{-}C-\underset{\underset{R'}{|}}{Ni(CO)_3} \xrightarrow{\text{solvent}} R\overset{\overset{O}{\|}}{-}C-R' + (\text{solvent})Ni(CO)_3
$$

## 2.10   Protection and activation by metals

Co-ordination of a metal atom to an alkene can occur by donation of electrons from the double bond to the metal atom, and normal alkene reactivity is thereby lost. This amounts to a protective function since the double bond can be restored by subsequent removal of the metal atom. In addition, co-ordination of an alkene, alkyne, or arene or of an appropriate carbanion or carbonium ion, to a transition metal atom can result in a dramatic change in the chemical reactivity of the organic moiety, and chemical transformations of such ligands can be brought about with retention of the metal–carbon bonds. For example, the highly reactive cyclopentadienyl anion, $C_5H_5^{\ominus}$, when $\pi$-bonded to a transition metal atom, can behave as a stable 'aromatic' system, and various electrophilic substitution reactions can be performed on the organic ligand which do not directly involve the metal.

e.g., $\qquad L_nM-C_5H_5 + X^{\oplus} \longrightarrow L_nM-C_5H_4X + H^{\oplus}$

where $X^{\oplus}$ is a suitable electrophilic species.

Such reactions can have useful consequences for organic synthesis since it is often possible at the end of the sequence of bond-making and bond-breaking processes to remove the deactivating ('protecting') or activating metal atom. Some examples are given below.

Nucleophilic substitution into a diene system can be effected by formation of the tricarbonyliron complex followed by hydride ion abstraction and attack on the resulting carbonium ion by a nucleophile such as water, cyanide ion or methoxide ion. [53, 54] The carbonium ion intermediate is stabilized by the iron tricarbonyl moiety, and in the last step the protecting metal atom is removed by oxidation.

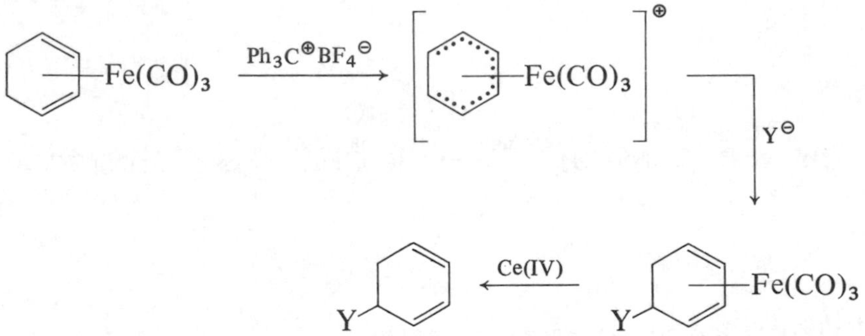

Formation of a π-bonded chlorobenzenetricarbonylchromium complex greatly enhances the ease of nucleophilic displacement of the chlorine atom from the aromatic ring, and the tricarbonylchromium function appears to act as a strong electron withdrawing group. [55]

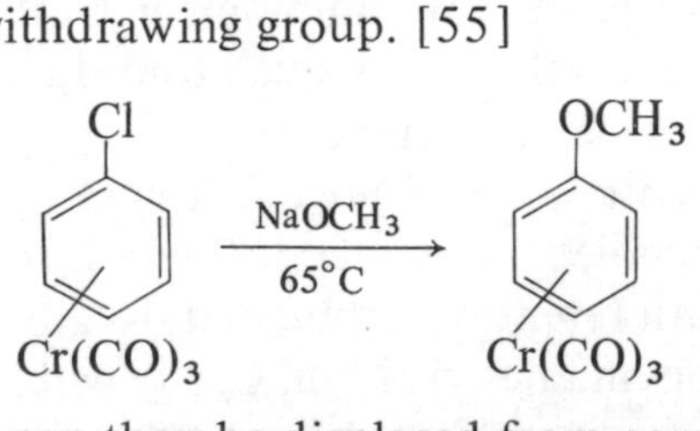

The organic ligand can then be displaced from combination with the chromium atom, e.g., by reaction with a more powerful nucleophile such as triphenylphosphine.

On the other hand, the benzene ring in a σ-bonded iron carbonyl complex is much more susceptible to electrophilic substitution than is benzene itself. Formylation can be effected by the Vilsmeier reagent (a mixture of $POCl_3$

and PhN(CHO)Me) in 80% yield, as compared with 2% for free benzene under identical conditions. [56] Activation of the ring is still observed even when the iron atom is separated from the benzene ring by a methylene group.

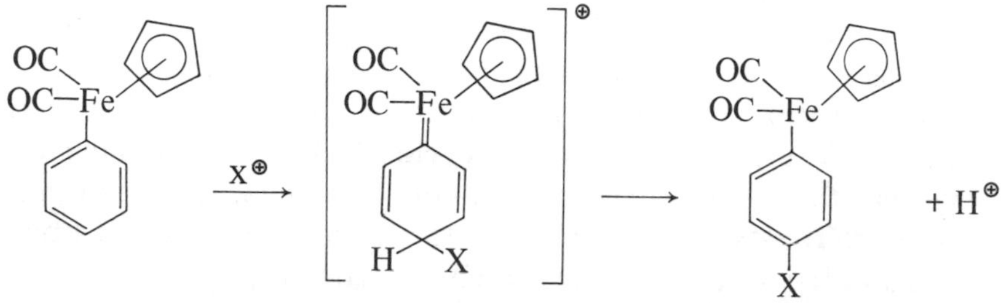

The reactivity of the $\pi$-bonded cyclo-octatetraenetricarbonyliron is also very different from that of the free tetraene. The complex behaves as an aromatic compound and can be formylated using a Vilsmeier reaction, the substituted tetraene being recovered from the complex by oxidation with a cerium(IV) salt. [57]

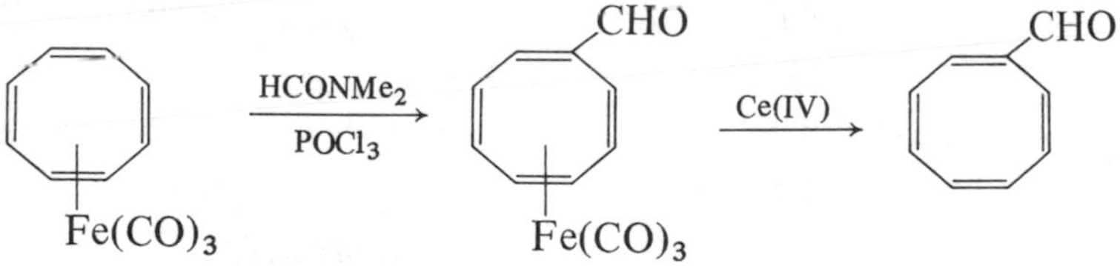

Specific protection of a carbon–carbon triple bond in an alkenylalkyne can be effected by octacarbonyldicobalt. [58]

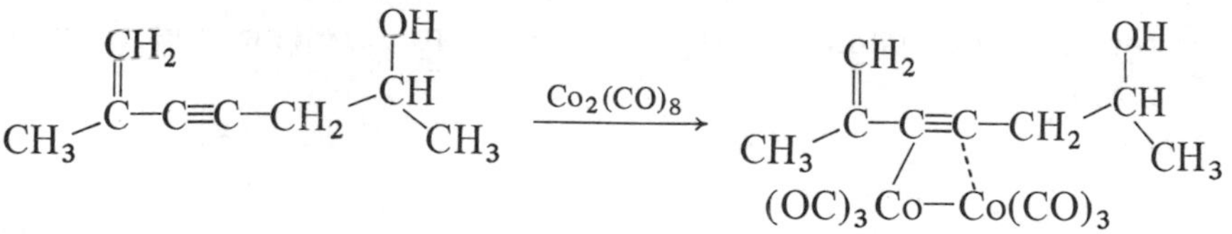

After formation of the cobalt complex, normal addition reactions of the carbon–carbon double bond can take place. The modified organic ligand still retaining a triple bond can be regenerated by oxidation of the complex with iron(III) nitrate. Normal Friedel–Crafts acylation of diphenylacetylene is unsuccessful; protection of the triple bond by reaction with octacarbonyl-dicobalt gives a complex which undergoes ready *para*-substitution in the

35

benzene rings, and from which the cobalt can again be removed with regeneration of the triple bond. [59]

Hydrogen cyanide does not normally add to an inactivated carbon–carbon double bond, but will add in the presence of octacarbonyldicobalt. [60] Presumably in this case the cobalt compound forms a complex with hydrogen cyanide which is a more powerful electrophile than is hydrogen cyanide.

Metal co-ordination can play an important role in stabilizing organic molecules which are themselves unstable, e.g. cyclobutadiene, $o$-quinodimethane, trimethylenemethane and norbornadien-7-one. The metal complex can serve as a source of the unstable organic molecule, which can be released from the complex in a variety of different environments. [61–65]

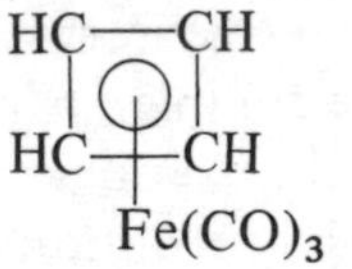

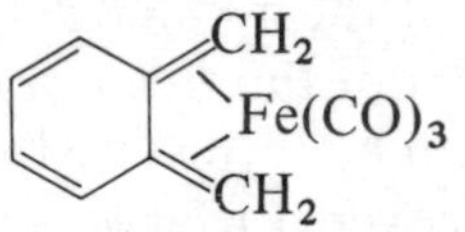

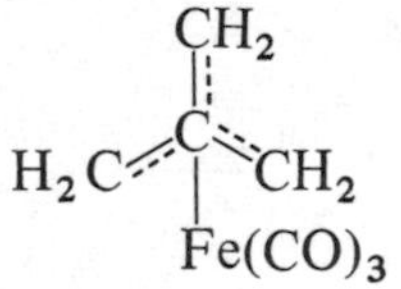

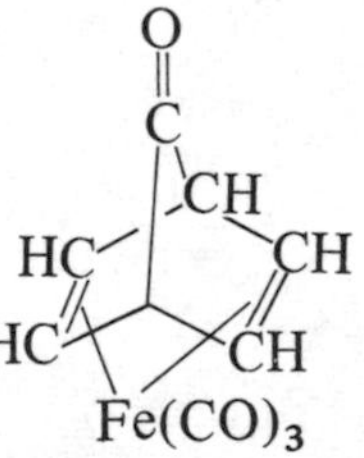

It should be noted that the metal complex is a unique compound in its own right. The free organic ligand is not present as such in the complex, and the binding forces in the ligand are very different in the presence and absence of the metal atom.

# Part II Synthesis of carbon–carbon bonds

# Reactions of carbanions 3

## 3.1 Introduction

The concept of carbanion, $R:^\ominus$, implies the existence of carbon acids in which a hydrogen atom attached to carbon is sufficiently acidic to be removed by a suitable base (B:).

$$-C\equiv C-H + B: \rightleftharpoons -C\equiv C:^\ominus + BH^\oplus \tag{3.1}$$

$$\text{>C=C<}_H + B: \rightleftharpoons \text{>C=C}^\ominus + BH^\oplus \tag{3.2}$$

$$\text{−C}_H + B: \rightleftharpoons \text{−C}^\ominus + BH^\oplus \tag{3.3}$$

The weak acidity of acetylene and its mono-substituted derivatives (Equation 3.1) is well known; these compounds readily form sodium, potassium, magnesium, copper(I) or silver salts by attack of a suitable anion on the hydrogen atom.

$$RC\equiv CH + NaNH_2 \longrightarrow RC\equiv C:^\ominus Na^\oplus + NH_3$$

$$RC\equiv CH + CH_3MgI \longrightarrow RC\equiv C-MgI + CH_4$$

A hydrogen atom attached to a vinyl carbon atom (Equation 3.2) is very much less acidic than hydrogen attached to an acetylenic carbon, and examples of direct ionization in the presence of a strong base are rare. However, vinyl (and aryl) organometallics can be prepared by indirect methods.

$$CH_2=CHCl \xrightarrow{Li + 2\% Na} CH_2=CHLi + LiCl$$

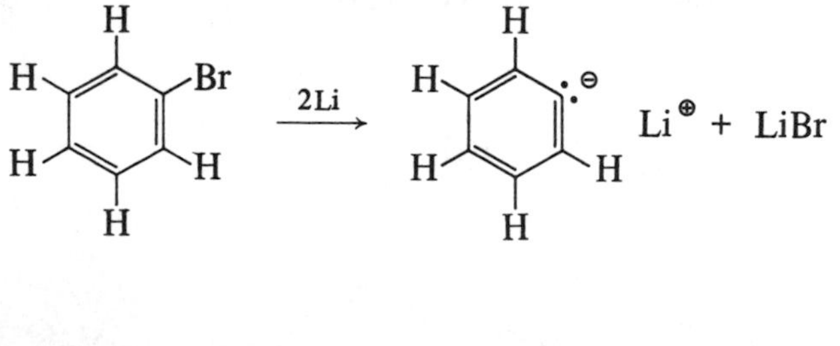

$$\xrightarrow{2Li} \quad Li^\oplus + LiBr$$

With simple alkanes, the attached hydrogen atoms are even less acidic, and direct ionization is virtually impossible. However, ease of ionization is related to the stability of the resulting anion, and any feature which allows delocalization of the negative charge over two or more atoms will promote anionic stability.

Allylic and benzylic carbanions are stabilized by delocalization of the negative charge over three and seven carbon atoms respectively.

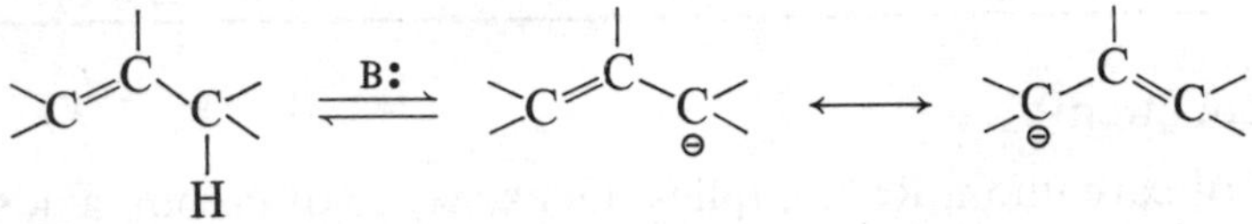

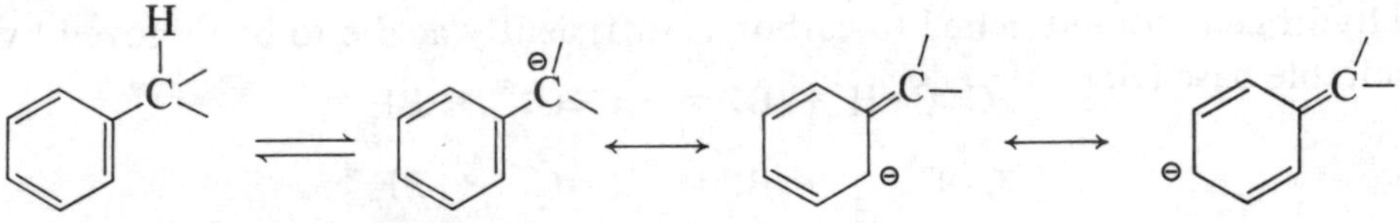

The stability of carbanions is also enhanced by delocalization of the negative charge from carbon to another more electronegative atom. A very large part of synthetic organic chemistry depends on the existence in organic molecules of functional groups which render such an ionization possible.

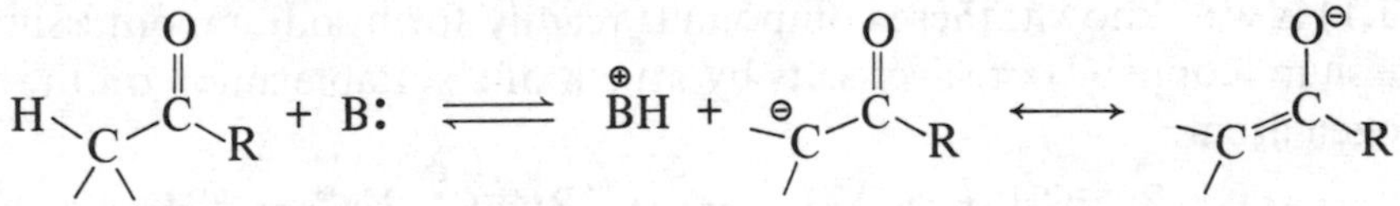

An adjacent carbonyl is the best known example of an acidifying functional group, and the carbanion chemistry of aldehydes, ketones and esters is very extensive. Many important, even classical, reactions such as the Claisen ester condensation, the aldol reaction, and acetoacetic and malonic ester alkylations fall into this class.

Other electron-withdrawing (acidifying) groups include nitro, cyano, sulphonyl and sulphinyl.

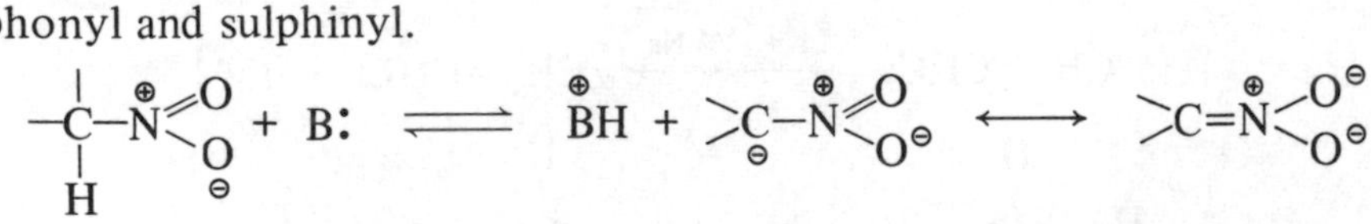

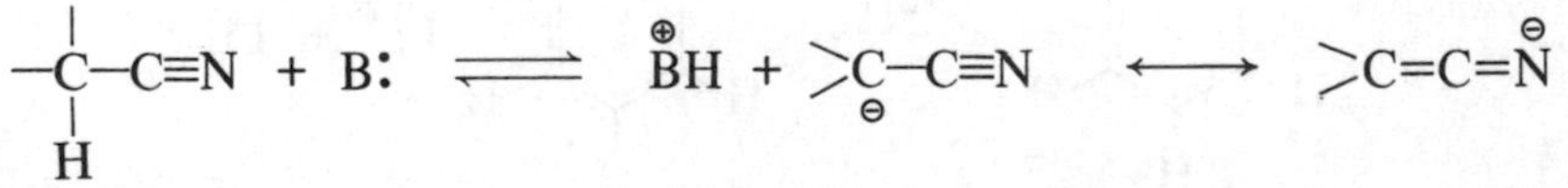

$$-\overset{\underset{|}{H}}{\underset{O}{C}}-\overset{\overset{O}{\|}}{S}-R + B\text{:} \rightleftharpoons \overset{\oplus}{B}H + \underset{O}{\overset{\overset{O}{\|}}{>}}\overset{\ominus}{C}-S-R \longleftrightarrow \underset{O}{\overset{O^{\ominus}}{>}}C=S-R$$

In alkylation reactions of the above anions, C-alkylation usually predominates but the ambident character of these anions sometimes results in O- or N-alkylation along with C-alkylation (cf. Section 2.2).

Carbanions can also be stabilized by delocalization of an electron pair in molecular orbitals on uncharged multiple chlorine or sulphur atoms, [66] or on a positively-charged phosphorus, sulphur, or nitrogen atom.

$$HCCl_3 + B\text{:} \rightleftharpoons \overset{\oplus}{B}H + \left[ \underset{Cl}{\overset{Cl}{C}}Cl \right]^{\ominus}$$

$$R-\underset{H}{\overset{S-CH_2}{C}}\underset{S-CH_2}{} + B\text{:} \rightleftharpoons \overset{\oplus}{B}H + \left[ R-\overset{\ddot{}}{C}\underset{S-CH_2}{\overset{S-CH_2}{}} \right]^{\ominus}$$

The carbanions generated from phosphonium, sulphonium, or ammonium salts form a special class known as ylids and are strictly outside the scope of this book, not being organometallic compounds. However, they are generated in many cases by use of an appropriate organometallic reagent, such as phenyllithium, to provide the necessary strong base:

$$\underset{R'}{\overset{R}{>}}H-\overset{\oplus}{C}-PPh_3 \rightleftharpoons^{B\text{:}} \underset{R'}{\overset{R}{>}}\overset{\ominus}{\text{:}C}-\overset{\oplus}{P}Ph_3 \longleftrightarrow \underset{R}{\overset{R}{>}}C=PPh_3$$

$$CH_3-\overset{\oplus}{S}\underset{CH_3}{\overset{CH_3}{<}} \rightleftharpoons^{B\text{:}} \overset{\ominus}{\text{:}}CH_2-\overset{\oplus}{S}\underset{CH_3}{\overset{CH_3}{<}} \longleftrightarrow CH_2=S\underset{CH_3}{\overset{CH_3}{<}}$$

There is a direct correlation between the reactivity of a metal alkyl or aryl compound and the ionic character of its carbon–metal bond. Ionic character is determined by a number of factors including size of the cation, its charge, the solvation of both cation and anion, and the degree of association of the organometallic compound in the solvent in question. The reactivity of organometallic compounds and the ionic nature of carbon–metal bonds decreases across the series $K > Na > Li > Mg > Zn > Cd > Hg$.

Organolithium compounds can be used in reactions with certain carbonyl groups where Grignard reagents are unreactive. On the other hand, organocadmium compounds are less reactive and therefore more selective than Grignard reagents. Differences such as these are often exploited in synthetic procedures.

## 3.2 Alkali–metal compounds

Sodium, potassium, rubidium and caesium salts of carbanions are essentially ionic and sodium alkyls are considered to be crystalline compounds with a high lattice energy. Ethylsodium can be shown to undergo dissociation in diethylzinc, according to the following equation.

$$NaEt + ZnEt_2 \rightleftharpoons \overset{\oplus}{Na} + \overset{\ominus}{ZnEt_3}$$

Most reactions of alkali–metal salts of carbanions are largely independent of the metal ion. An early and important example of such a reaction is alkylation of the carbanion derived from a terminal alkyne, $RC{\equiv}CH$, by means of an alkyl halide, $R'X$:

$$RC{\equiv}\overset{\ominus}{C}\!: M^{\oplus} + R'X \longrightarrow RC{\equiv}CR' + M^{\oplus}X^{\ominus}$$

The two best known reactions of metal salts of ketocarbanions are probably alkylation with an alkyl halide, and acylation with an acid chloride:

42

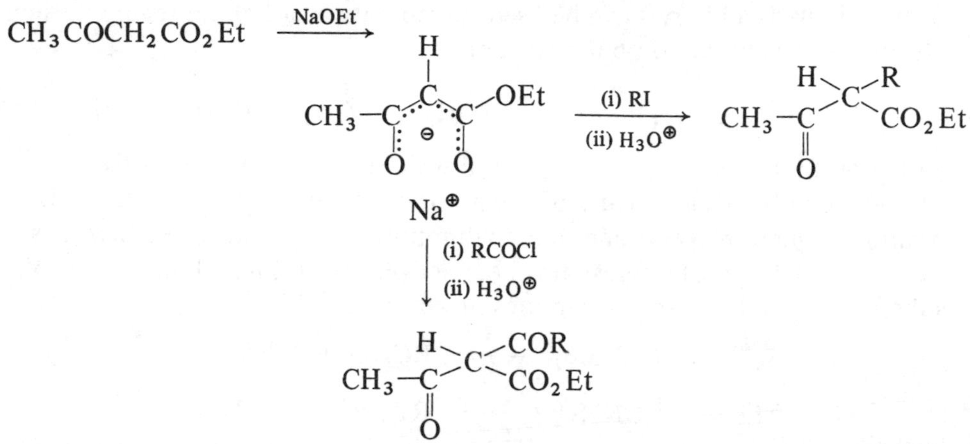

$$CH_3COCH_2CO_2Et \xrightarrow{\text{NaOEt}}$$

$$\xrightarrow[\text{(ii) } H_3O^\oplus]{\text{(i) RI}}$$

(i) RCOCl
(ii) $H_3O^\oplus$

The solvent can play a critical role in the reactions of carbanions due to solvation both of the carbanion itself and the associated metal cation. This can be illustrated by the results of methylation of 2-methylcyclohexanone using either diethyl ether or liquid ammonia as solvent.

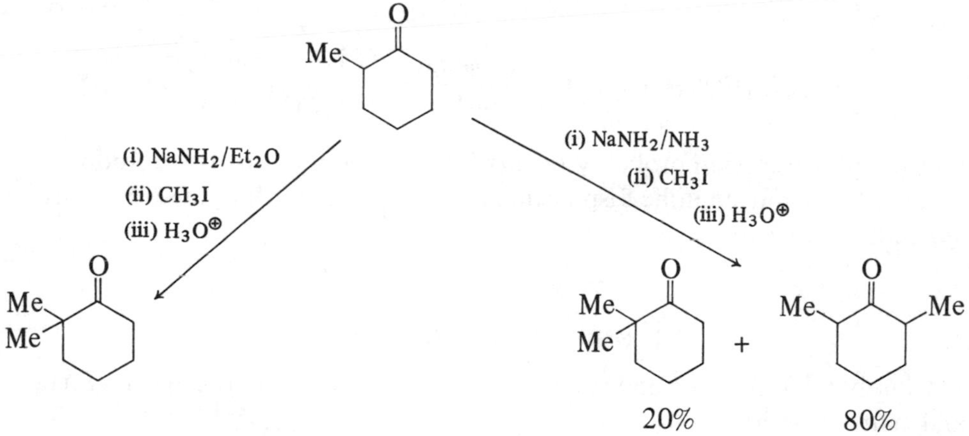

(i) NaNH$_2$/Et$_2$O
(ii) CH$_3$I
(iii) H$_3$O$^\oplus$

(i) NaNH$_2$/NH$_3$
(ii) CH$_3$I
(iii) H$_3$O$^\oplus$

20%        80%

Lithium alkyls are essentially covalent; ethyllithium is soluble in benzene and exists as a hexameric aggregate in this solvent. With such compounds it is possible to carry out carbanion reactions in non-polar media.

$$C_2H_5Li + CO_2 \xrightarrow{\text{benzene}} C_2H_5COOLi$$

43

Lithium alkyls and aryls can be used to generate a carbanion from another molecule, especially in non-polar solvents.

$$C_6H_5Li + (CH_3)_4\overset{\oplus}{P}\overset{\ominus}{I} \longrightarrow (CH_3)_3\overset{\oplus}{P}-\overset{\ominus}{C}H_2 + C_6H_6 + LiI$$

Organolithium compounds have found wide utility in the synthesis of carbon–carbon bonds and many of their reactions parallel those of the corresponding Grignard reagents and organoaluminium compounds (see Sections 3.3 and 3.4). Of special interest are the reactions of lithium alkyls with $N,N$-disubstituted amides to give aldehydes or ketones. [67]

$$RLi + HCONMe_2 \longrightarrow RCHO + LiNMe_2$$

$$RLi + R'CONMe_2 \longrightarrow RCOR' + LiNMe_2$$

Symmetrical ketones can also be formed by reaction of lithium alkyls with carbon monoxide [68] (cf. Chapter 5):

$$2RLi + 3CO \longrightarrow RCOR + 2LiCO$$

A further carbanion reaction of some generality is oxidative dimerization by means of iodine.

$$2CH_3COCH_2CO_2Et \xrightarrow[\text{(ii) }I_2]{\text{(i) NaOEt}} \begin{array}{l} CH_3COCHCO_2Et \\ \quad\quad\quad | \\ CH_3COCHCO_2Et \end{array}$$

This coupling reaction probably occurs by formation of an alkyl iodide followed by a nucleophilic displacement reaction of this iodide with a further molecule of carbanion:

$$R^{\ominus} + I_2 \longrightarrow RI + I^{\ominus}$$

$$RI + R^{\ominus} \longrightarrow R-R + I^{\ominus}$$

Closely related to the above oxidative dimerizations are the Wurtz–Fittig and Wurtz reactions.

A lithium or sodium aryl is the likely intermediate in the first of these reactions, in which an aryl halide and alkyl halide are coupled in the presence of lithium or sodium metal.

$$C_6H_5Br + 2Na \longrightarrow C_6H_5Na + NaBr$$

$$C_6H_5Na + RBr \longrightarrow C_6H_5R + NaBr$$

The first step, formation of phenylsodium from bromobenzene and sodium metal, is a consequence of electron donation from sodium to the aromatic nucleus.

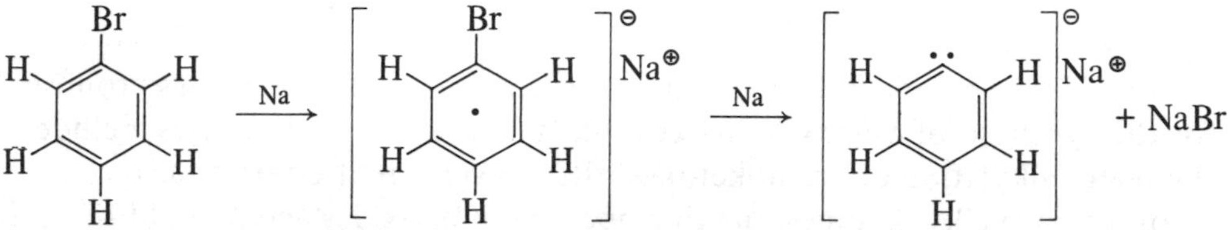

The second step is then a normal nucleophilic displacement on the alkyl halide by the aryl carbanion (cf. Section 2.2).

For the Wurtz reaction, the coupling of two molecules of alkyl halide in the presence of sodium, there is a good deal more controversy as to the mechanism, but it seems probable that free radicals are involved (cf. Section 2.4). It has been suggested that a sodium alkyl is formed initially and that the sodium ions in this ionic aggregate coordinate with further molecules of alkyl halide:

$$RX \xrightarrow{2Na} R^{\ominus}Na^{\oplus} \xrightarrow{R'X} R^{\ominus}[R'-X \rightarrow Na]^{\oplus}$$

Electron exchange then occurs between unsolvated carbanion and the complex cation to yield two radicals, $R\cdot$ and $[R'-X \rightarrow Na]\cdot$, the latter then undergoing further decomposition:

$$R^{\ominus}[R'-X \rightarrow Na]^{\oplus} \longrightarrow R\cdot + [R'-X \rightarrow Na]\cdot \longrightarrow R'\cdot + NaX$$

The radicals $R\cdot$ and $R'\cdot$ can then combine to give the symmetrical and un-symmetrical hydrocarbons R—R, R'—R', and R—R'; radical disproportionation to RH, R'H and the related alkenes can also occur.

Yields in the Wurtz reaction can be high in cases where the radical coupling process is intramolecular.

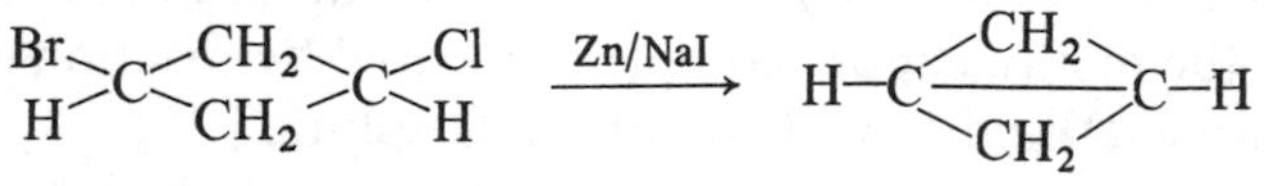

## 3.3    Organocopper compounds

A modern variant of the Wurtz reaction is found in the use of lithium di-methylcopper for coupling reactions with alkyl, aryl, or vinyl halides. [69]

The reagent is made by reaction of methyllithium with copper(I).

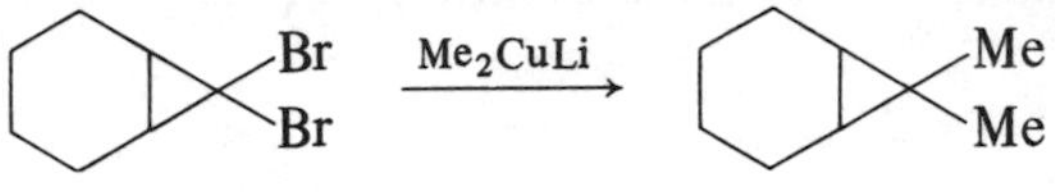

Other lithium copper dialkyls can be used and the reaction shows promise for the synthesis of otherwise inaccessible hydrocarbons. Other uses include the ready alkylation of bromoketones, the conversion of esters or acid chlorides into alkyl ketones, the ring opening of epoxides and 1,4-additions to $\alpha,\beta$-unsaturated ketones. [70–74]

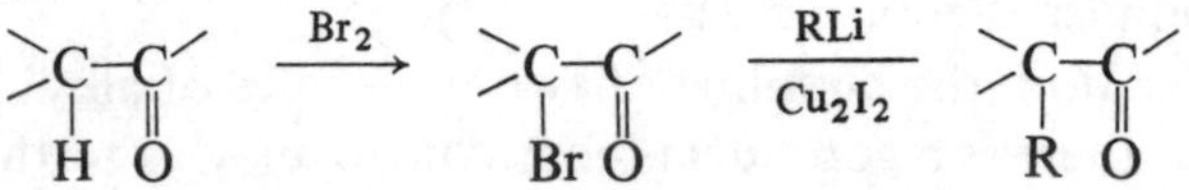

$$RCO_2Et \quad \text{or} \quad RCOCl \xrightarrow{Me_2CuLi} RCOCH_3$$

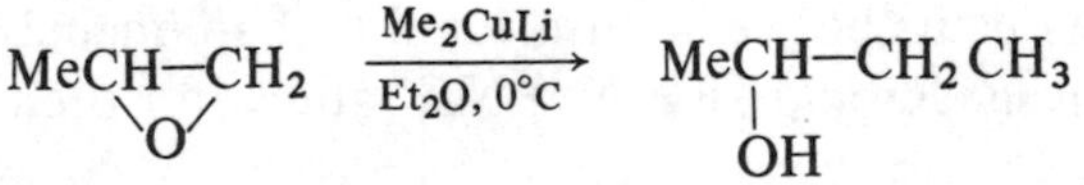

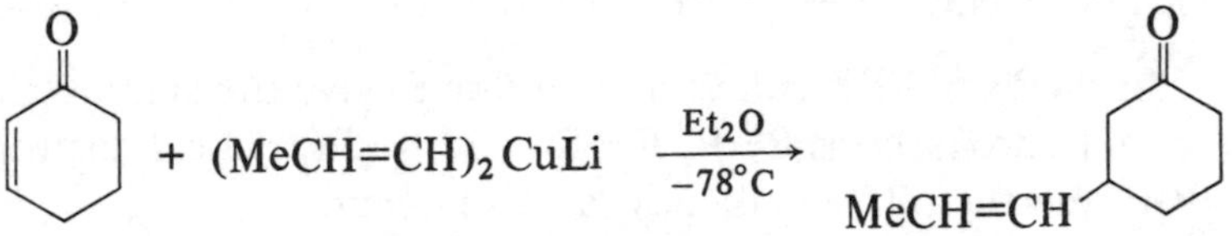

Conjugate addition of a methyl group to an $\alpha,\beta$-unsaturated ketone can also be effected by the methylcopper complex prepared by reaction of methyl-lithium with copper(I) iodide and trimethylphosphite. [75]

Whereas lithium dimethylcopper effects the conversion of RX into $RCH_3$, the more hindered lithium dibutylcopper brings about halogen–copper exchange instead, even with vinyl halides, e.g.,

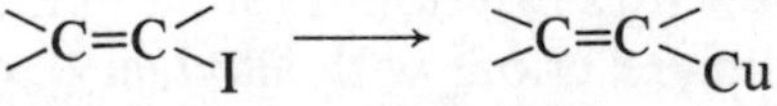

The reagent can therefore induce cyclization of vinyl halides having a suitably placed carbonyl group. [76]

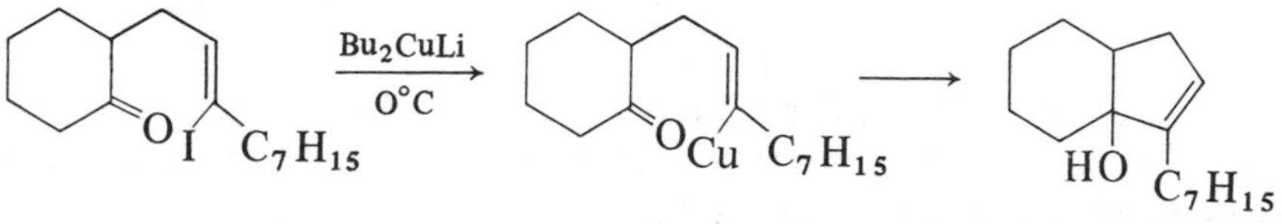

## 3.4 Organomagnesium compounds

The organomagnesium Grignard reagents, commonly shown as RMgX, are usually prepared by reaction of the corresponding alkyl or aryl halides with metallic magnesium in ether solution. Benzene or toluene can be used as solvent provided a complexing ether or tertiary amine is present. As a result of a reaction at the metal surface, a magnesium atom becomes 'inserted' into a carbon–halogen bond. It is possible that a bond forms between the halogen atom and a magnesium atom at the surface of the solid metal, followed by homolytic cleavage of the carbon–halogen bond and recombination of the organic radical with the metal atom which is now attached to halogen.

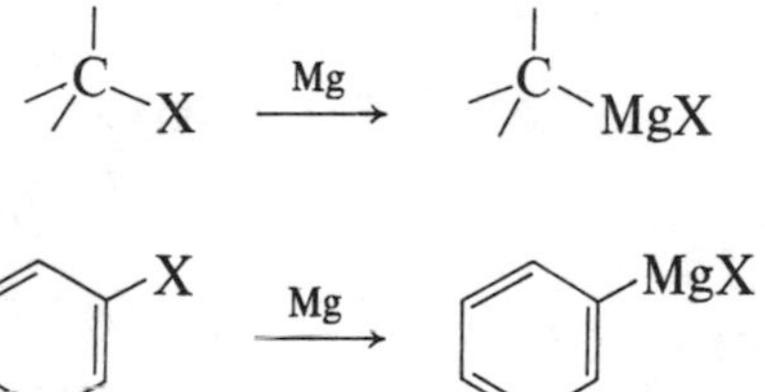

The order of halide reactivity is $I > Br > Cl$ and the alkyl halide may be primary, secondary, or tertiary. In the case of aryl chlorides, and vinyl halides in general, tetrahydrofuran or another higher-boiling solvent is required, rather than the usual diethyl ether. Alkynyl magnesium reagents cannot be prepared from the corresponding halide, but are obtained by direct metallation of the terminal alkyne (cf. Section 3.1). 1,2-Dihalogenoalkanes undergo an elimination reaction with metallic magnesium to form the corresponding alkene and magnesium halide.

$$BrCH_2CH_2Br \xrightarrow{\phantom{xx}Mg\phantom{xx}} [Br-CH_2CH_2-MgBr] \longrightarrow CH_2=CH_2 + MgBr_2$$

47

The Grignard reagent should strictly be represented as $R{-}\overset{\displaystyle S}{\underset{\displaystyle S}{Mg}}{-}X$,

where S represents an ether (or tertiary amine) molecule. In ether solution the complex is in equilibrium with the corresponding dialkyl (or diaryl) magnesium compound and solvated magnesium halide:

$$R{-}\overset{OEt_2}{\underset{OEt_2}{Mg}}{-}X \rightleftharpoons R{-}\overset{OEt_2}{\underset{OEt_2}{Mg}}{-}R + X{-}\overset{OEt_2}{\underset{OEt_2}{Mg}}{-}X$$

Grignard reagents have found wide application in the synthesis of carbon-carbon bonds. A summary of the more important reactions is given in the diagram, where the group R initially attached to the magnesium atom can be alkyl, aryl, alkenyl, or alkynyl.

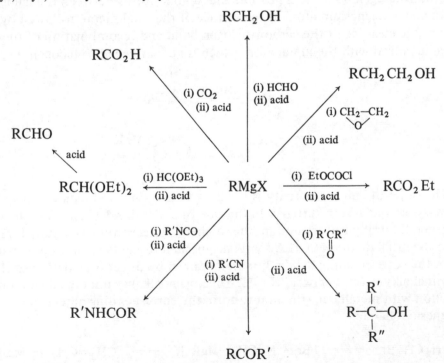

48

Some specific examples are listed below. [77] In each case, the initial product is a magnesium alkoxide or related complex, from which the final organic product is obtained by the addition of aqueous acid.

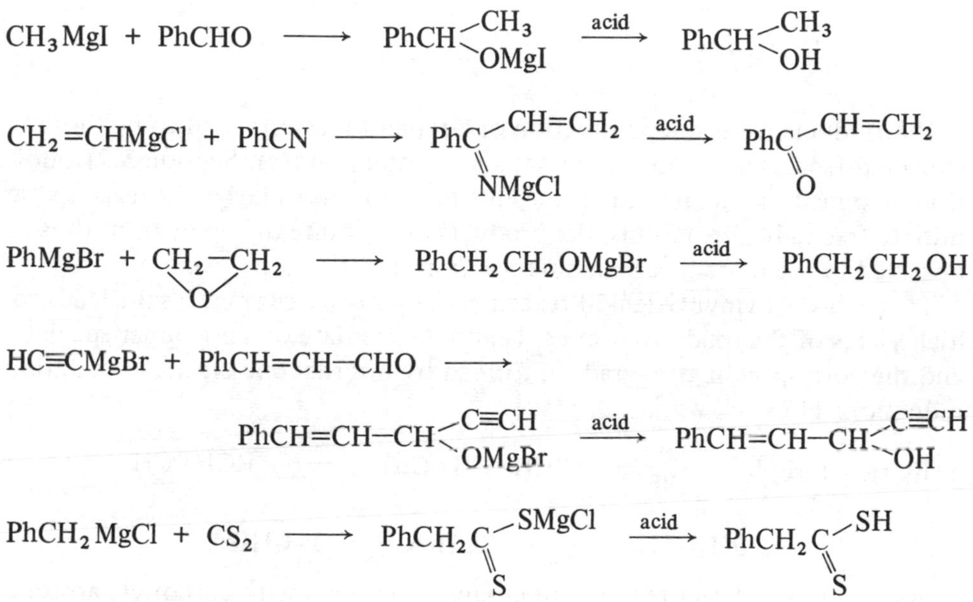

$$CH_3MgI + PhCHO \longrightarrow PhCH\underset{OMgI}{\overset{CH_3}{<}} \xrightarrow{acid} PhCH\underset{OH}{\overset{CH_3}{<}}$$

$$CH_2=CHMgCl + PhCN \longrightarrow PhC\underset{\overset{\|}{NMgCl}}{\overset{CH=CH_2}{-}} \xrightarrow{acid} PhC\underset{\overset{\|}{O}}{\overset{CH=CH_2}{-}}$$

$$PhMgBr + \underset{O}{CH_2-CH_2} \longrightarrow PhCH_2CH_2OMgBr \xrightarrow{acid} PhCH_2CH_2OH$$

$$HC\equiv CMgBr + PhCH=CH-CHO \longrightarrow$$

$$PhCH=CH-CH\underset{OMgBr}{\overset{C\equiv CH}{<}} \xrightarrow{acid} PhCH=CH-CH\underset{OH}{\overset{C\equiv CH}{<}}$$

$$PhCH_2MgCl + CS_2 \longrightarrow PhCH_2C\underset{\overset{\|}{S}}{\overset{SMgCl}{-}} \xrightarrow{acid} PhCH_2C\underset{\overset{\|}{S}}{\overset{SH}{-}}$$

All of the above reactions can be rationalized by supposing that the group initially attached to magnesium behaves as a carbanion (cf. Section 2.2). The general reaction with a carbonyl group can therefore be shown:

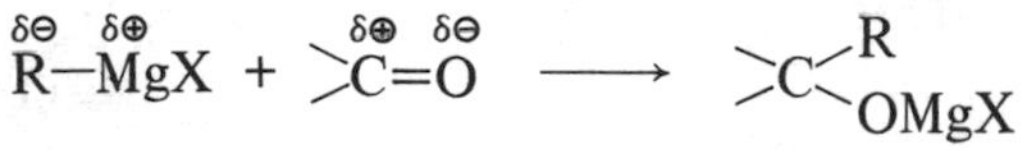

$$\overset{\delta\ominus\ \delta\oplus}{R-MgX} + \overset{\delta\oplus\ \delta\ominus}{>C=O} \longrightarrow >C\underset{OMgX}{\overset{R}{<}}$$

In the reaction of RMgX with a ketone $R_2'CO$ the ketone is thought to displace one of the co-ordinated ether molecules; this complex then reacts with a second molecule of Grignard reagent. [78–80]

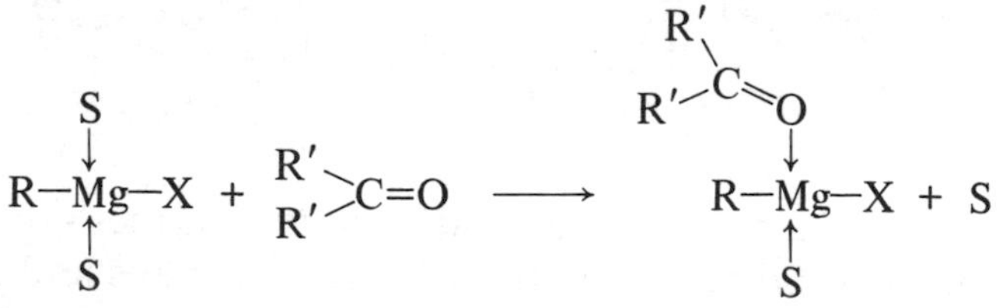

$$\underset{\overset{\uparrow}{S}}{\overset{\overset{S}{\downarrow}}{R-Mg-X}} + \underset{R'}{\overset{R'}{>}}C=O \longrightarrow \underset{\overset{\uparrow}{S}}{\overset{\overset{R'-C=O}{\overset{\downarrow}{}}}{R-Mg-X}} + S$$

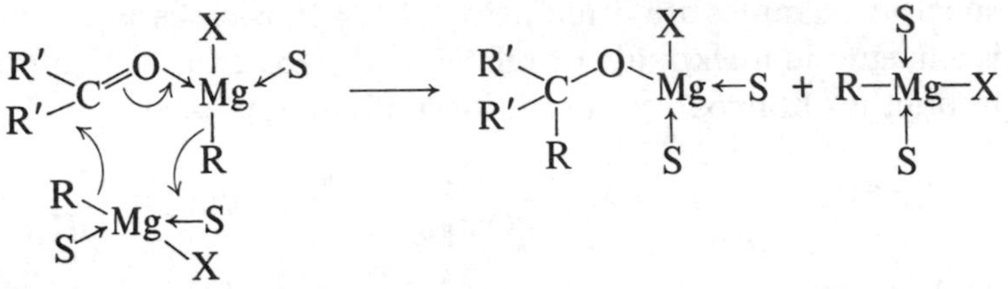

By-products from the preparation of Grignard reagents tend to be those expected from Wurtz-type couplings and eliminations (cf. Section 3.2); addition of transition-metal oxidizing agents to solutions of Grignard reagents can initiate free radical reactions, the products being quite different from those observed in the normal carbanionic reactions (cf. Section 2.4).

In the case of vinyl Grignard reagents, coupling by copper(I) salts leads to high yields of the oxidative dimer via an intermediate organocopper species and the corresponding free radical formed by internal one-electron oxidation-reduction: [17]

$$RCH{=}CHMgBr \xrightarrow[THF]{Cu_2Cl_2} [RCH{=}CH{-}Cu] \longrightarrow RCH{=}CH\cdot + Cu$$

$$2RCH{=}CH\cdot \longrightarrow RCH{=}CH{-}CH{=}CHR$$

As well as the direct reaction of Grignard reagents with carbonyl carbon atoms, 1,4-addition to $\alpha,\beta$-unsaturated carbonyl compounds is also possible. [81]

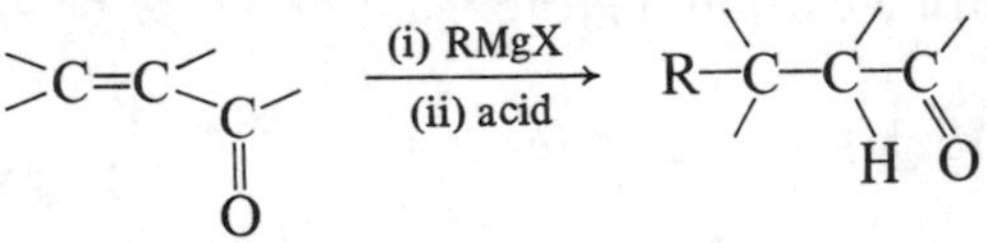

This type of addition can be catalysed by copper(I) salts, and presumably organocopper intermediates are again involved (cf. Section 3.3):

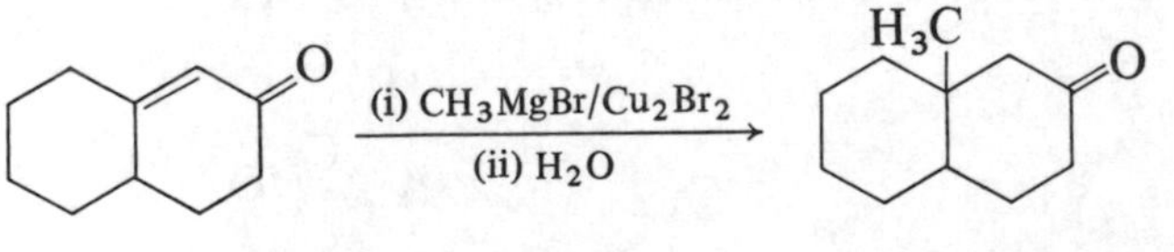

Some important Grignard reactions leading to the synthesis of carbon–carbon bonds have been discussed in this Chapter; reactions leading to the

50

synthesis of carbon–hydrogen bonds are discussed in Section 6.2, and of carbon–nitrogen, carbon–phosphorus, carbon–oxygen, carbon–sulphur and carbon–halogen bonds in Section 7.2.

## 3.5    Organoaluminium compounds

Trimethylaluminium, a commercially available chemical, has been shown to be a valuable reagent for the preparation of methyl ketones from acid chlorides, and could displace dimethylcadmium for this purpose. [82]

$$3RCOCl + Me_3Al \longrightarrow 3RCOMe + AlCl_3$$

In a related reaction, triethylaluminium reacts with nitriles to yield ethyl ketones.

$$RC{\equiv}N \xrightarrow{Et_3Al} RC\!\!\underset{NAlEt_2}{\overset{Et}{\diagdown}} \xrightarrow{H_2O} RC\!\!\underset{O}{\overset{Et}{\diagdown}}$$

Yields are highest when the ratio of triethylaluminium to nitrile is $2:1$, and the reaction may well involve a six-membered transition state as shown below. However a four-membered transition state is also possible.

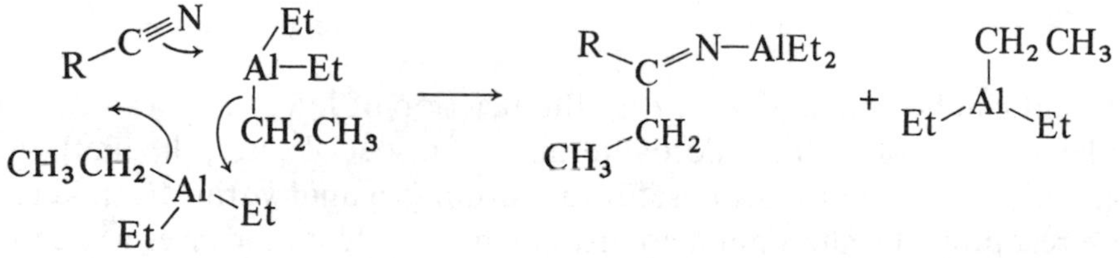

$$R\text{--}C\overset{\equiv N}{\underset{}{}}\quad \overset{Et}{\underset{CH_2CH_3}{Al\text{--}Et}}$$

$$CH_3CH_2\text{--}Al\overset{}{\underset{Et}{\diagdown}}Et$$

$$\longrightarrow \quad R\text{--}\underset{CH_3\diagup CH_2}{C}\text{=}N\text{--}AlEt_2 \quad + \quad \overset{CH_2CH_3}{\underset{Et}{Al}}Et$$

Trimethylaluminium reacts with tertiary alcohols, and hence also with ketones, to provide a new synthesis of hydrocarbons.

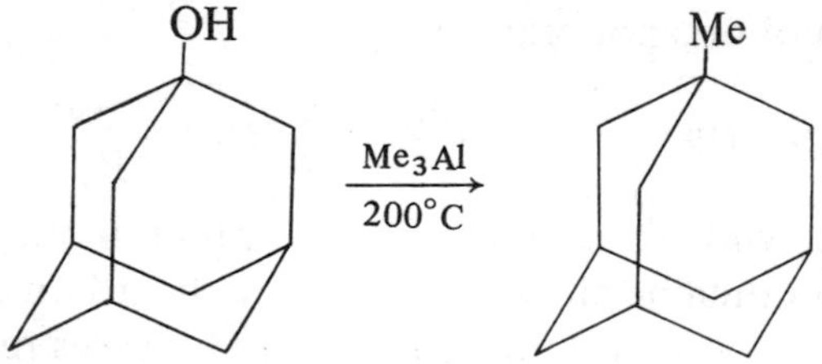

$$\xrightarrow[200°C]{Me_3Al}$$

51

$$\underset{R'}{\overset{R}{>}}C=O \xrightarrow{Me_3Al} \left[ \underset{R'}{\overset{R}{>}}C\overset{Me}{\underset{OAlMe_2}{<}} \right] \longrightarrow \underset{R'}{\overset{R}{>}}C\overset{Me}{\underset{Me}{<}}$$

This latter reaction, surprisingly, is catalysed by the addition of traces of water or a carboxylic acid. When a carboxylic acid is heated with excess trimethylaluminium and a trace of water the carboxyl group is converted into a tertiary butyl group by a similar series of reactions. [83]

$$RCO_2H \xrightarrow{Me_3Al} RCMe_3$$

In all these cases the trialkylaluminium clearly acts as a source of alkyl carbanions.

The higher alkyl derivatives of aluminium, especially triethylaluminium, are discussed in Section 4.4.3 in connection with the polymerization of alkenes, and in Sections 7.2.3 and 4.1.1 in relation to the industrial synthesis of long-chain primary alcohols and α-olefins respectively.

Diethylcyanoaluminium is a useful reagent for the conjugate addition of hydrogen cyanide to α,β-unsaturated ketones. [84–86]

$$\underset{R}{\overset{R}{>}}C=C\overset{R}{\underset{\underset{O}{\overset{||}{C-R}}}{<}} \xrightarrow{Et_2AlCN} \underset{R}{\overset{R}{>}}C\overset{R}{\underset{CN}{\overset{|}{-}}}C\overset{H}{\underset{\underset{O}{\overset{||}{C-R}}}{<}}$$

The reagent can be prepared *in situ* by the reaction of hydrogen cyanide with triethylaluminium or diethylchloroaluminium. In special cases, the diethylcyanoaluminium reagent is also useful for forming cyanohydrins from ketones. In these reactions the aluminium co-ordinated cyanide ion behaves as a more powerful carbanion than do the ethyl groups. Diisobutylhydridoaluminium is a useful reducing agent and is capable of effecting hydroalumination of alkynes which undergo *cis* addition to yield vinyl alanes (cf. Section 6.3 for a related application of organoboranes).

$$RC\equiv CR \xrightarrow{(C_4H_9)_2AlH} \underset{R}{\overset{H}{>}}C=C\overset{Al(C_4H_9)_2}{\underset{R}{<}}$$

Treatment of the vinyl alane with methyllithium in ether yields an aluminate complex, which can then be allowed to react with carbon dioxide or cyanogen to yield sterically pure α,β-unsaturated acids and nitriles respectively. [87, 88]

52

$$\begin{array}{c} H \\ R \end{array} C=C \begin{array}{c} Al(C_4H_9)_2 \\ R \end{array} \xrightarrow{CH_3Li} \left[ \begin{array}{c} CH_3 \\ | \\ H \\ R \end{array} C=C \begin{array}{c} Al(C_4H_9)_2 \\ R \end{array} \right]^{\ominus} Li^{\oplus}$$

$$\xleftarrow[\text{(ii) } H_3O^{\oplus}]{\text{(i) } CO_2} \qquad \xrightarrow{NC-CN}$$

$$\begin{array}{c} H \\ R \end{array} C=C \begin{array}{c} CO_2H \\ R \end{array} \qquad \begin{array}{c} H \\ R \end{array} C=C \begin{array}{c} CN \\ R \end{array}$$

It is noteworthy that the isomeric lithium aluminate complex can be formed by the addition of lithium diisobutylmethylhydridoaluminate to the alkyne; this addition occurs in *trans* fashion. Thus the isomeric α,β-unsaturated acids and nitriles can also be obtained. [89]

$$RC{\equiv}CR \xrightarrow{\quad Li \left[ \begin{array}{c} C_4H_9 \\ C_4H_9 \end{array} Al \begin{array}{c} CH_3 \\ H \end{array} \right]^{\ominus} \quad} \left[ \begin{array}{c} H \\ R \end{array} C=C \begin{array}{c} R \\ Al(C_4H_9)_2 \\ | \\ CH_3 \end{array} \right]^{\ominus} Li^{\oplus}$$

$$\xleftarrow[\text{(ii) } H_3O^{\oplus}]{\text{(i) } CO_2} \qquad \xrightarrow{NC-CN}$$

$$\begin{array}{c} H \\ R \end{array} C=C \begin{array}{c} R \\ CO_2H \end{array} \qquad \begin{array}{c} H \\ R \end{array} C=C \begin{array}{c} R \\ CN \end{array}$$

In the above carbonation and cyanation processes the lithium aluminate complexes exhibit carbanionic behaviour, in much the same way as do vinyl lithium and vinyl magnesium compounds.

## 3.6    Organoboron compounds

α-Halogeno esters, ketones, and nitriles undergo ready alkylation by trialkyl-boranes in the presence of suitable bases such as potassium t-butoxide, or preferably the hindered base potassium 2,6-di(t-butyl)phenoxide, [90] e.g.

53

$$R_3B + BrCH_2CO_2Et \xrightarrow[\text{Me}_3\text{COH}]{\text{Me}_3\text{COK}} RCH_2CO_2Et$$

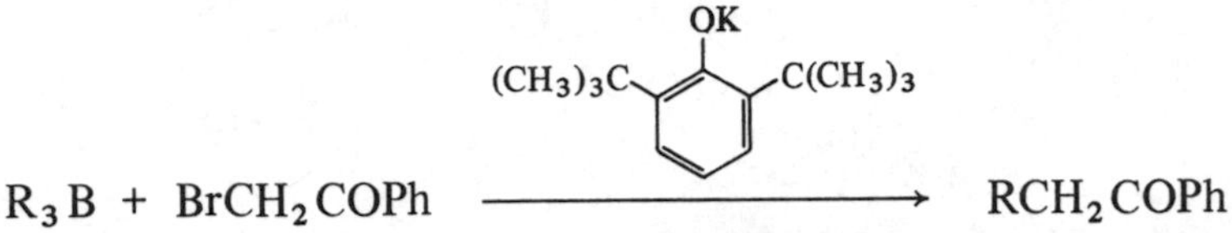

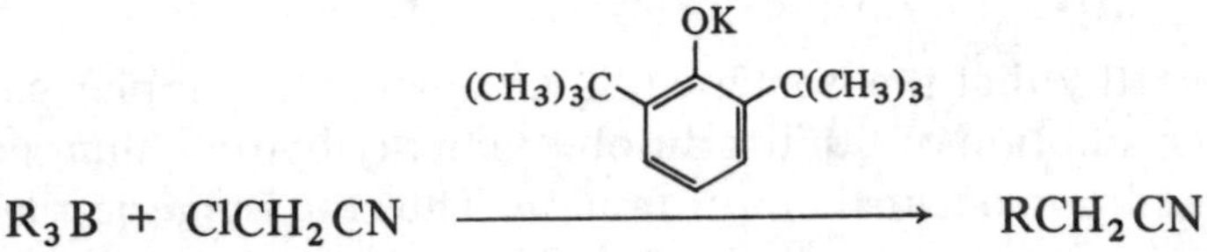

$$R_3B + BrCH_2COPh \xrightarrow{\text{(see structure)}} RCH_2COPh$$

where the reagent above the arrow is:

$$\underset{\text{(CH}_3)_3\text{C}}{\overset{\text{OK}}{\bigcirc}}\text{C(CH}_3)_3$$

$$R_3B + ClCH_2CN \xrightarrow{\text{(see structure)}} RCH_2CN$$

where the reagent above the arrow is:

$$\underset{\text{(CH}_3)_3\text{C}}{\overset{\text{OK}}{\bigcirc}}\text{C(CH}_3)_3$$

The mechanism of this reaction probably involves attack of a carbanion on the trialkylborane followed by a carbanion migration from boron to carbon.

$$BrCH_2CO_2Et \xrightarrow{\text{Me}_3\text{COK}} Br\overset{\ominus}{C}HCO_2Et$$

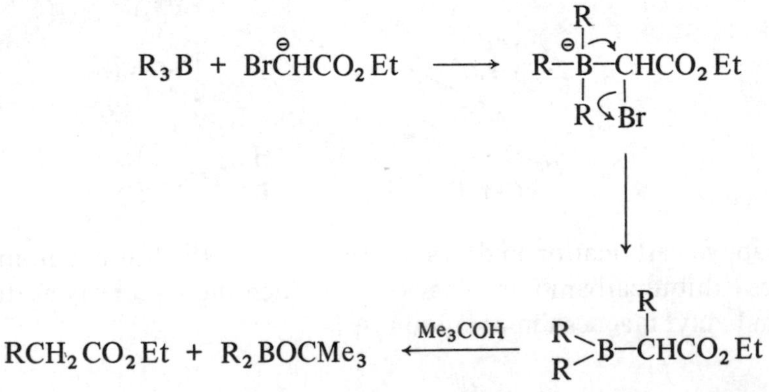

$$R_3B + Br\overset{\ominus}{C}HCO_2Et \longrightarrow \underset{\underset{R}{|} \underset{Br}{|}}{\overset{R}{\underset{|}{R-\overset{\ominus}{B}-CHCO_2Et}}}$$

$$\downarrow$$

$$RCH_2CO_2Et + R_2BOCMe_3 \xleftarrow{\text{Me}_3\text{COH}} \underset{R}{\overset{R}{\underset{}{R}}>B-\overset{R}{\overset{|}{C}HCO_2Et}$$

The use of 9-borabicyclononane allows for a more efficient conversion of an alkene into its alkylated product (see Chapter 5.5). α-Halogenoboron compounds can also be formed by the photobromination of trialkylboranes and in this case water is a strong enough base to allow group migration to occur.

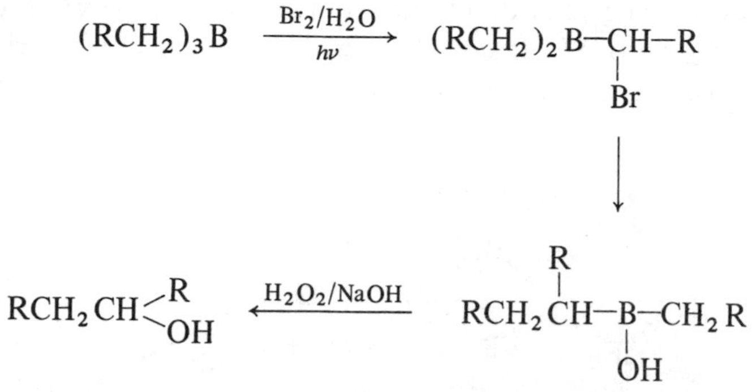

$$(RCH_2)_3B \xrightarrow[h\nu]{Br_2/H_2O} (RCH_2)_2B-\underset{\underset{\displaystyle Br}{|}}{CH}-R$$

$$RCH_2CH\underset{OH}{\overset{R}{<}} \xleftarrow{H_2O_2/NaOH} RCH_2\underset{\underset{\displaystyle OH}{|}}{CH}-\overset{\overset{\displaystyle R}{|}}{B}-CH_2R$$

Trialkylboranes undergo similar reactions with carbanions derived from diazoketones and also with carbanions of the ylid type, e.g.

$$R_3B + N_2CHCOCH_3 \xrightarrow{THF} RCH_2COCH_3$$

$$R_3B + \overset{\ominus}{C}H_2\overset{\oplus}{S}Me_2 \xrightarrow[\text{(ii) }H_2O_2/NaOH]{\text{(i) DMSO}/-10°} ROH + RCH_2OH$$

A recent discovery is the synthesis of cyclopropanes from 3-chloroalkenes via an organoborane; treatment of this intermediate with alkali gives the cyclopropane presumably by a similar carbanionic mechanism as shown below:

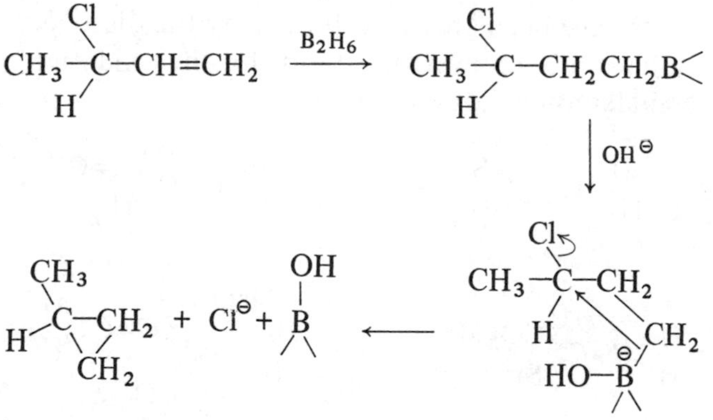

$$CH_3-\underset{H}{\overset{Cl}{\underset{/}{\overset{\backslash}{C}}}}-CH=CH_2 \xrightarrow{B_2H_6} CH_3-\underset{H}{\overset{Cl}{\underset{/}{\overset{\backslash}{C}}}}-CH_2CH_2B{<}$$

$$\downarrow OH^{\ominus}$$

$$\underset{CH_2}{\overset{CH_3}{\underset{\backslash/}{\overset{|}{C}}}}{-}CH_2 \; H{-}\overset{}{} + Cl^{\ominus} + \underset{/\backslash}{B} \longleftarrow CH_3-\underset{H}{\overset{Cl}{\underset{/}{\overset{\searrow}{C}}}}{-}\underset{\underset{\displaystyle HO-\underset{/\backslash}{B}}{}}{\overset{CH_2}{\underset{\ominus}{\diagup}}}CH_2$$

## 3.7    Organonickel compounds

A major new discovery in synthesis is that of Corey and co-workers who have found that allyl bromides readily form $\pi$-complexes with tetracarbonylnickel in benzene; subsequent reaction of the complex with an alkyl or aryl iodide leads to a novel form of Wurtz coupling, which proceeds by a carbanionic rather than a free radical mechanism. [91]

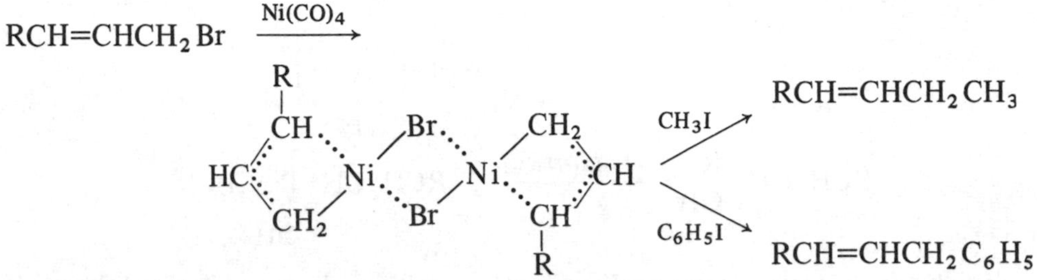

$$RCH{=}CHCH_2\,Br \xrightarrow{\quad Ni(CO)_4 \quad}$$

In the case of allylic dihalides, the reaction provides an efficient route to large ring alkadienes.

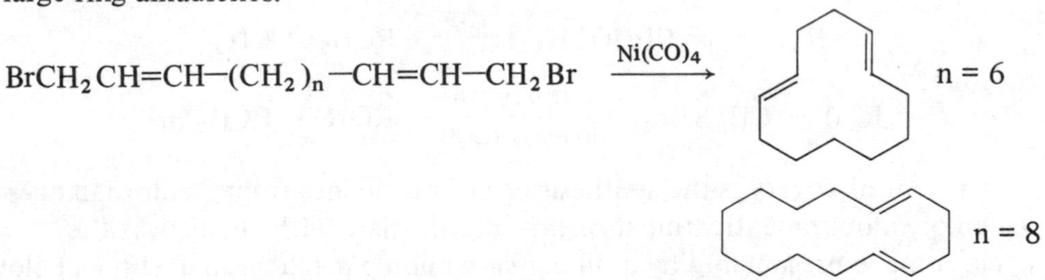

$$BrCH_2\,CH{=}CH{-}(CH_2)_n{-}CH{=}CH{-}CH_2\,Br \xrightarrow{\quad Ni(CO)_4 \quad}$$

n = 6

n = 8

These reactions have been greatly extended by the discovery that bis(1,5-cyclo-octadiene)nickel is a powerful catalyst for the oxidative coupling of allyl, benzyl, aryl, and alkenyl halides. [92]

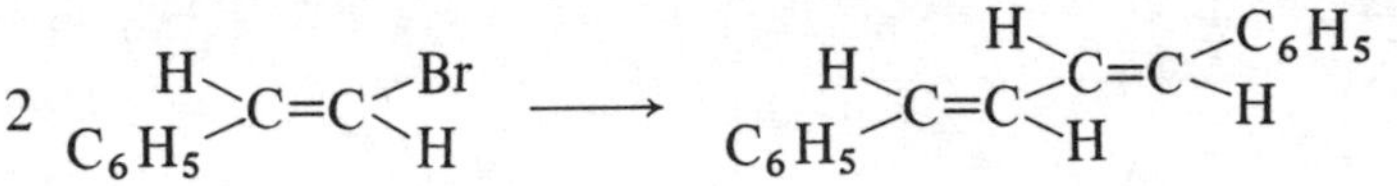

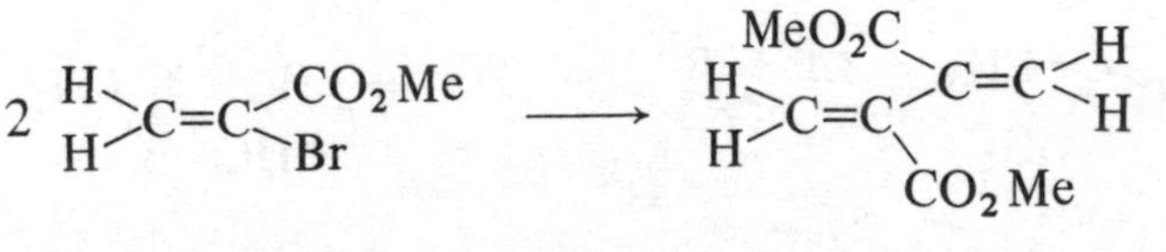

## 3.8 Carbanionic reactions involving other metals

In general, organozinc, cadmium, and to a lesser extent mercury and tin compounds are comparable to organomagnesium compounds in chemical

56

behaviour, but are much less reactive (cf. Chapter 2.2). Most reactions of these organometallic compounds can be generalized in terms of carbanion-like intermediates, and the lesser reactivity can be attributed to a higher degree of covalency in the carbon–metal bond. The following reactions of organo-mercury compounds illustrate these points. [93, 94]

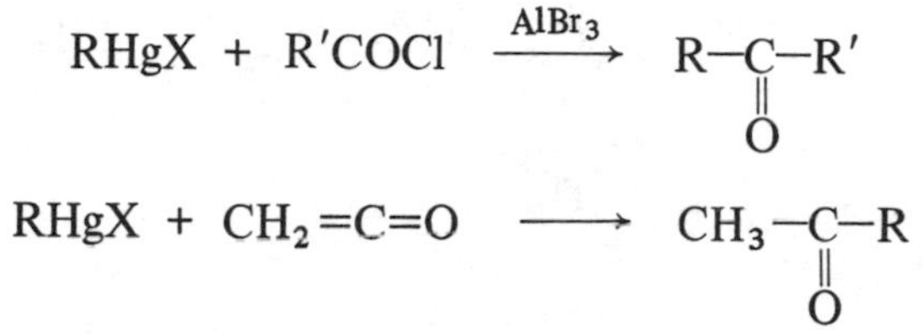

$$RHgX + R'COCl \xrightarrow{AlBr_3} R-\underset{\underset{O}{\|}}{C}-R'$$

$$RHgX + CH_2{=}C{=}O \longrightarrow CH_3-\underset{\underset{O}{\|}}{C}-R$$

$\alpha$-Functionally substituted organotin compounds $R_3SnCH_2X$ such as organotin nitriles, esters, ketones or amides, undergo addition to carbonyl compounds in a manner analogous to Grignard reagents, to afford a variety of $\beta$-hydroxy nitriles, -esters, -ketones and -amides. [95]

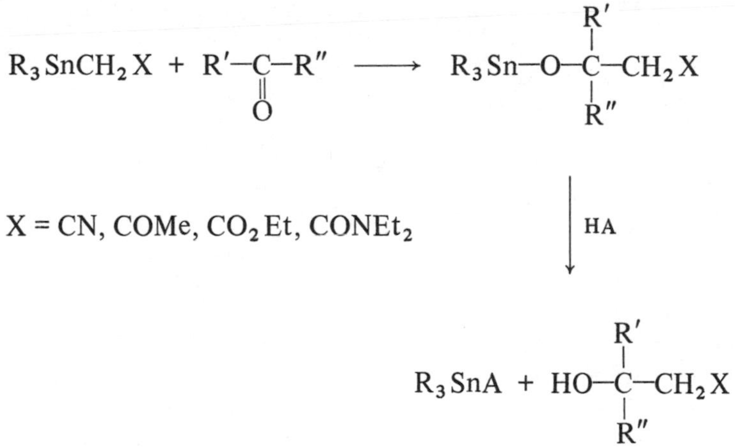

$$R_3SnCH_2X + R'-\underset{\underset{O}{\|}}{C}-R'' \longrightarrow R_3Sn-O-\underset{\underset{R''}{|}}{\overset{\overset{R'}{|}}{C}}-CH_2X$$

$$X = CN, COMe, CO_2Et, CONEt_2 \qquad \Big\downarrow HA$$

$$R_3SnA + HO-\underset{\underset{R''}{|}}{\overset{\overset{R'}{|}}{C}}-CH_2X$$

In some well-known examples involving less-common metals the metal compound is not organometallic in the sense defined in this book but is rather a substance in which the metal is linked covalently to one or more oxygen atoms. This is true of the zinc compound formed as an intermediate in the Reformatsky reaction (cf. Section 1.1) and of the very useful thallium(I) derivatives of 1,3-dicarbonyl compounds which undergo alkylation on the central carbon atom in virtually quantitative yield. [96]

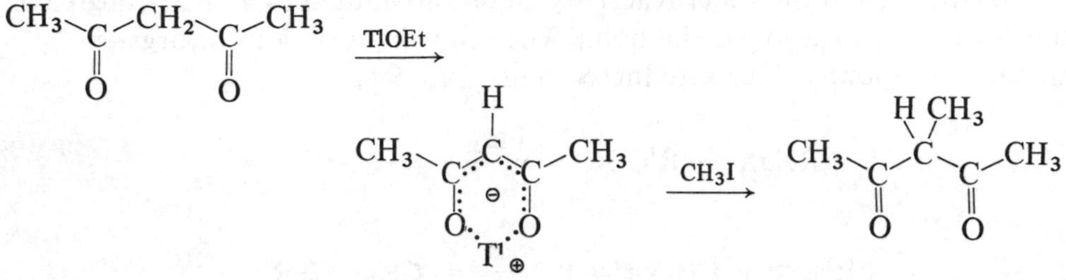

# Alkenes, alkynes and arenes: synthesis, isomerization, and polymerization

# 4

## 4.1  Synthesis

### 4.1.1  *Alkenes and alkynes*

Treatment of ethers with very strong bases such as alkylsodium or alkyl-lithium compounds can yield alkenes, but side reactions usually occur.

$$\underset{\underset{H \quad OR}{|\quad\ |}}{>C\!-\!C<} \ +\ R'Na \ \longrightarrow\ >C\!=\!C< \ +\ RONa \ +\ R'H$$

The pyrolysis of alkyllithium compounds having a $\beta$-hydrogen atom gives rise to the corresponding alkene and lithium hydride; [33] similar reactions are reported for alkylsodium, alkylpotassium, and alkylaluminium compounds. [97] This last reaction is of industrial importance in the synthesis of long-chain terminal alkenes (cf. Section 4.4.3).

$$\underset{\underset{H \quad Li}{|\quad|}}{>C\!-\!C<} \ \overset{\Delta}{\longrightarrow}\ >C\!=\!C< \ +\ LiH$$

$$>Al(CH_2)_n R \ \overset{\Delta}{\longrightarrow}\ \underset{H}{\overset{H}{>}}C\!=\!C\underset{(CH_2)_{n-2}R}{\overset{H}{<}} \ +\ >AlH$$

Grignard reagents can be pyrolysed by heating them in nonsolvating high-boiling solvents such as cumene. [98] Ethyltrimethyltin eliminates ethene on treatment with triphenylmethyl fluoroborate. [99]

$$CH_3CH_2\!-\!SnMe_3 \ +\ Ph_3\overset{\oplus}{C}\overset{\ominus}{B}F_4 \ \longrightarrow$$

$$CH_2\!=\!CH_2 \ +\ Ph_3CH \ +\ Me_3Sn^\oplus BF_4^\ominus$$

Other alkenes can be formed in similar reactions involving mercury and lead alkyls. In general, hydride abstraction can occur readily from the $\beta$-carbon atom of an alkyl group attached to a metal atom, provided that the resulting 'carbonium ion' intermediate can be stabilized by $\pi$-conjugation to the metal. Loss of the cationic metal group then gives rise to the product alkene.

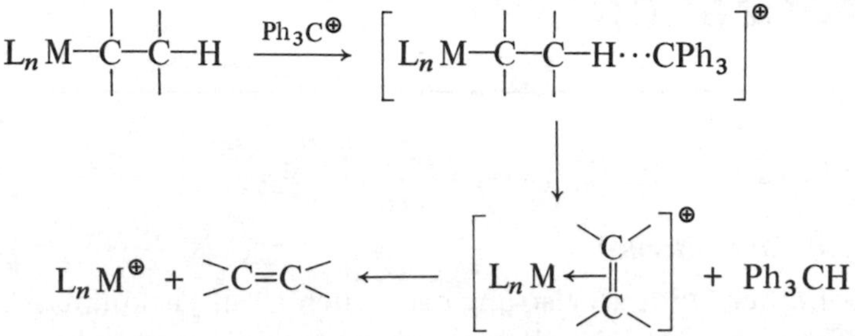

$$L_n M-\overset{|}{\underset{|}{C}}-\overset{|}{\underset{|}{C}}-H \xrightarrow{Ph_3C^\oplus} \left[L_n M-\overset{|}{\underset{|}{C}}-\overset{|}{\underset{|}{C}}-H\cdots CPh_3\right]^\oplus$$

$$L_n M^\oplus + {>}C{=}C{<} \longleftarrow \left[L_n M\leftarrow \overset{\diagup C\diagdown}{\underset{\diagdown C\diagup}{\|}}\right]^\oplus + Ph_3CH$$

These and the following reactions are examples of elimination processes in which the cationic leaving group contains a metal atom (cf. Section 2.7).

The dehalogenation of vicinal dihalides and related compounds can be accomplished in good yield by a variety of reagents, including metallic zinc and magnesium. [100] Where metals are employed, organometallic intermediates are highly probable.

$$BrCH_2CH_2Br \xrightarrow{Mg} [BrCH_2CH_2MgBr] \xrightarrow{elimination} CH_2{=}CH_2 + MgBr_2$$

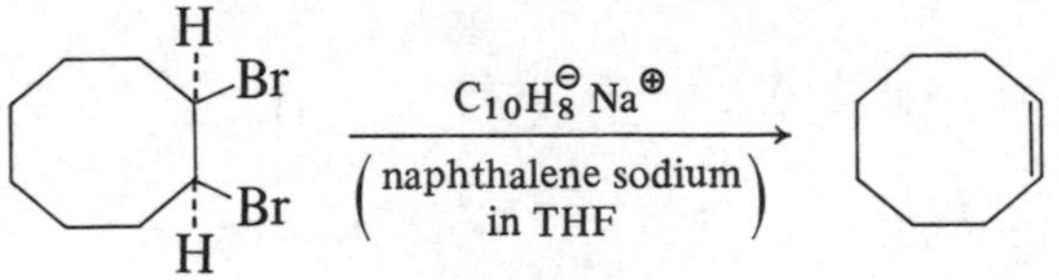

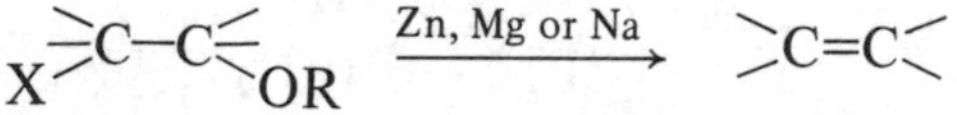

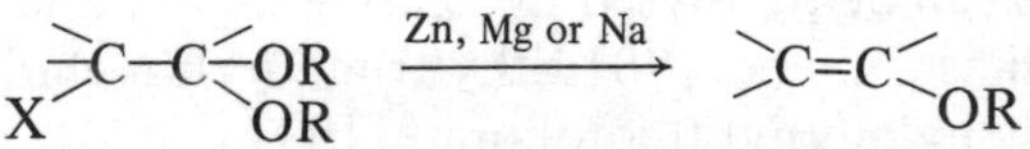

$$\underset{H}{\overset{H}{\text{(cyclooctane with Br, Br)}}} \xrightarrow[\left(\substack{\text{naphthalene sodium} \\ \text{in THF}}\right)]{C_{10}H_8^\ominus Na^\oplus} \text{(cyclooctene)}$$

$$\underset{X}{\overset{>}{}}C-C\underset{OR}{\overset{<}{}} \xrightarrow{Zn,\ Mg\ or\ Na} {>}C{=}C{<}$$

$$\underset{X}{\overset{>}{}}C-C\underset{OR}{\overset{OR}{}} \xrightarrow{Zn,\ Mg\ or\ Na} {>}C{=}C\underset{OR}{<}$$

$\beta$-Halogenoethyl organosilicon compounds undergo thermal fragmentation to afford alkenes, the trimethylsilyl moiety being an effective leaving group [101] (cf. Section 2.7).

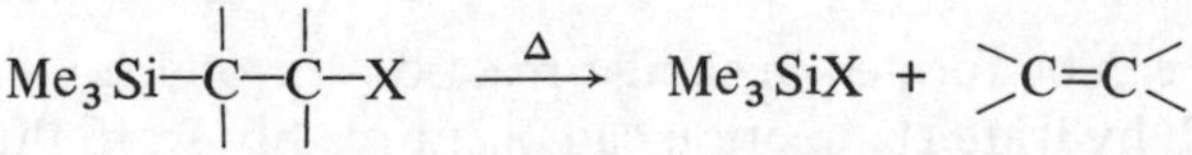

$$Me_3Si-\overset{|}{\underset{|}{C}}-\overset{|}{\underset{|}{C}}-X \xrightarrow{\Delta} Me_3SiX + {>}C{=}C{<}$$

An interesting preparation of 1-butene involves the acid hydrolysis of 2-butenylbromomercury. This organometallic compound undergoes reaction with hydrogen chloride about $10^7$ times faster than $n$-butylbromomercury, and since the product is more than 99% 1-butene rather than 2-butene, the following mechanism is proposed.

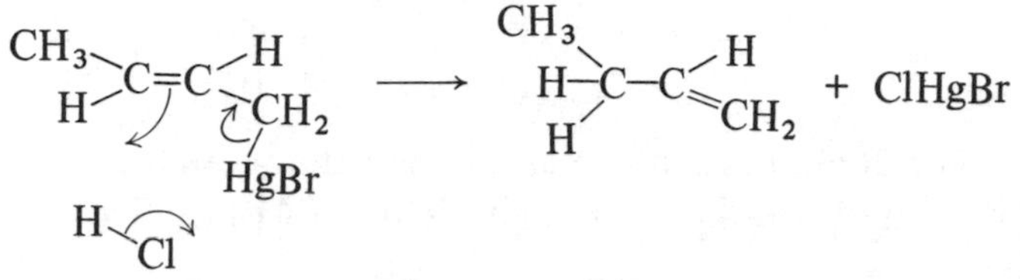

The reaction of $\beta$-halogenoethers with zinc, magnesium or sodium is known as the Boord reaction; this reaction also yields an alkene from compounds

$$X-\overset{|}{\underset{|}{C}}-\overset{|}{\underset{|}{C}}-Y \text{ where X is halogen and Y is } -OCOR, -OSO_2R, -NR_2, \text{ or } -SR.$$

These elimination reactions ensure that the desired carbon–carbon double bond is produced in a specific position; this is also the case for the synthesis of ketenes from $\alpha$-halogeno acyl halides.

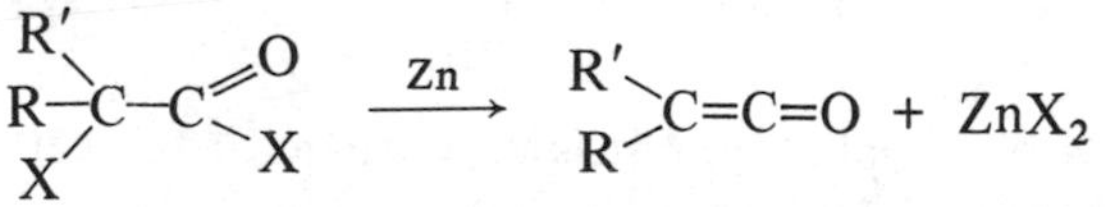

Perhaps the most versatile synthesis of alkenes is provided by the Wittig reaction, which is outside the scope of this book. However, a somewhat related process involves reaction of an aldehyde or ketone with the Grignard reagent derived from methylene bromide. [102]

$$CH_2Br_2 \xrightarrow{Mg/Hg} CH_2(MgBr)_2$$

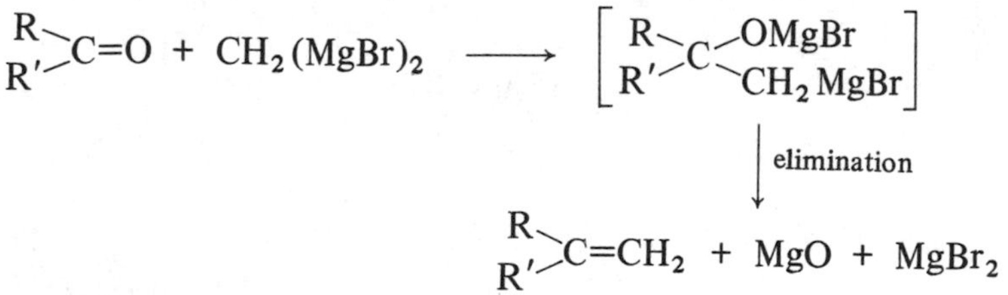

A standard method for the synthesis of alkenes is the partial reduction of alkynes. [103] Stereospecific *cis* reduction can be achieved via an intermediate organoborane adduct (see also Section 6.3).

$$RC{\equiv}CR' + R''_2BH \longrightarrow \underset{H}{\overset{R}{\diagdown}}C{=}C\underset{BR''_2}{\overset{R'}{\diagup}} \xrightarrow{\text{AcOH}}$$

$$\underset{H}{\overset{R}{\diagdown}}C{=}C\underset{H}{\overset{R'}{\diagup}} + R''_2BOAc$$

Diisobutylhydridoaluminium can also effect the reduction of disubstituted alkynes to the related *cis*-alkenes in high yield [104] (see Section 3.5).

$$RC{\equiv}CR \xrightarrow[45°C]{(C_4H_9)_2AlH} \underset{H}{\overset{R}{\diagdown}}C{=}C\underset{H}{\overset{R}{\diagup}}$$

A range of useful alkenes can be derived from the additions of diborane and alkyl boranes to alkynes. Hydroboration of a non–terminal alkyne yields a trivinyl borane which can be converted into a *cis*-alkene by treatment with acetic acid (see Section 6.3).

$$RC{\equiv}CR \xrightarrow{B_2H_6} \left(\underset{H}{\overset{R}{\diagdown}}C{=}C\underset{B}{\overset{R}{\diagup}}\right)_3 \xrightarrow{CH_3CO_2H} \underset{H}{\overset{R}{\diagdown}}C{=}C\underset{H}{\overset{R}{\diagup}}$$

In the case of a terminal alkyne it is very difficult to stop the hydroboration at the vinyl borane stage and consequently terminal alkenes cannot be prepared reliably by this method. The use of the sterically-hindered bis(3-methyl-2-butyl)borane makes possible the preparation of a monovinyl borane, from which the terminal alkene can be produced by hydrolysis.

$$RC{\equiv}CH + \left(\underset{Me_2CH-CH}{\overset{Me}{|}}\right)_2 BH \longrightarrow \underset{H}{\overset{R}{\diagdown}}C{=}C\underset{B}{\overset{H}{\diagup}}\left(\underset{Me}{\overset{CH-CHMe_2}{|}}\right)_2$$

$$\xrightarrow{CH_3CO_2H}$$

$$RCH{=}CH_2 + CH_3CO_2B\left(\underset{Me}{\overset{CH-CHMe_2}{|}}\right)_2$$

Vinyl boranes can also be oxidized by iodine in sodium hydroxide to a substituted alkene product. [105]

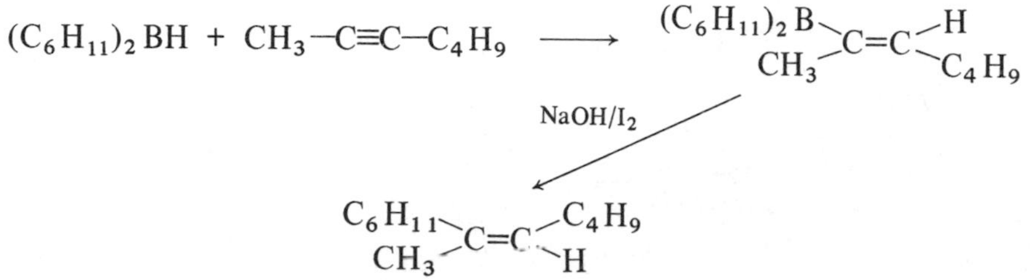

$$(C_6H_{11})_2 BH + CH_3-C\equiv C-C_4H_9 \longrightarrow \underset{CH_3}{\overset{(C_6H_{11})_2 B}{\phantom{x}}}C=C\underset{C_4H_9}{\overset{H}{\phantom{x}}}$$

NaOH/I$_2$

$$\underset{CH_3}{\overset{C_6H_{11}}{\phantom{x}}}C=C\underset{H}{\overset{C_4H_9}{\phantom{x}}}$$

This process involves migration of the original borane substituent on to the alkene carbon atom initially attached to boron. Geometrical isomerization occurs during oxidative removal of the boron and the result is stereospecific formation of an alkene in which the two substituents of the initial alkyne are now *trans*.

The synthesis of a vinyl ether from a metal carbene complex was outlined in Chapter 2.3.

The synthesis of a tetra-arylethene shown below may well involve an intermediate iron–carbene complex. [106]

$$2Ar_2CBr_2 + 2Fe(CO)_5 \longrightarrow Ar_2C=CAr_2 + 2FeBr_2 + 10CO$$

The classical synthesis of alkynes from tetrahalogenoalkanes and metallic zinc must certainly involve intermediate organozinc compounds.

$$\underset{Cl}{\overset{Cl}{\phantom{x}}}C-C\underset{Cl}{\overset{Cl}{\phantom{x}}} + 2Zn \longrightarrow -C\equiv C- + 2ZnCl_2$$

In general, however, there appear to be no really useful methods for the synthesis of alkynes from organometallic compounds.

## 4.1.2 *Arenes and arynes*

A classic reaction of organic chemistry is the synthesis of benzene by the trimerization of acetylene in a hot tube. Metal-catalysed versions of this thermal reaction are now well known and are discussed in Section 4.4.2 below.

Arynes are well recognized intermediates in certain reactions of 1,2-disubstituted arenes. Appropriate elimination of the *ortho* substituents

generates the aryne which can then dimerize, trimerize, or undergo nucleophilic or other addition reactions.

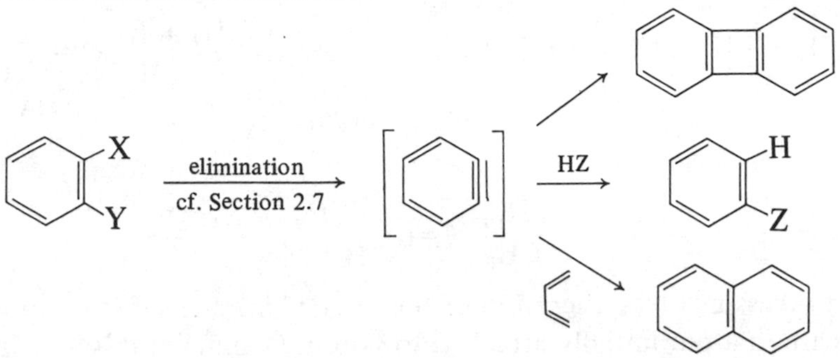

Many successful routes to arynes involve organometallic intermediates, e.g.

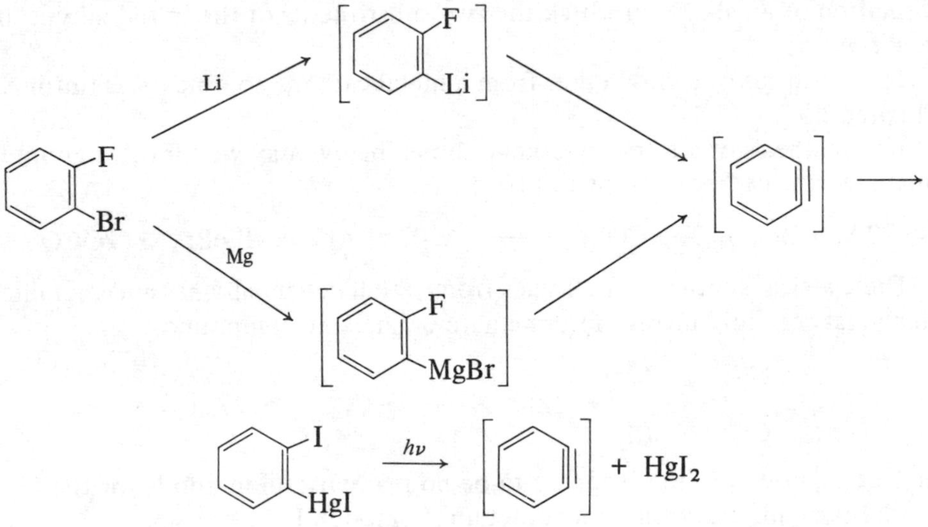

## 4.2    Isomerization

Certain metal salts and organometallic complexes can act as catalysts for the
isomerization of strained cycloalkanes, of alkenes, and in particular, of
alkadienes. Reaction conditions are usually mild and superior in many known
cases to those involving acid or base catalysis or thermal treatment. In addition,
many quite new reactions have been observed, and organometallic intermediates, either $\sigma$- or $\pi$-bonded, are clearly involved.

64

### 4.2.1  *Strained cycloalkanes*

These reactions are currently of great theoretical interest, but of rather little utility as yet in general synthesis. All the examples discussed below involve the breaking of one or more strained σ-bonds with formation of an isomeric alkene. The conversion of cubane, $C_8H_8$ into an isomeric '*cis*'-diene can be effected with a rhodium(I) catalyst. [107]

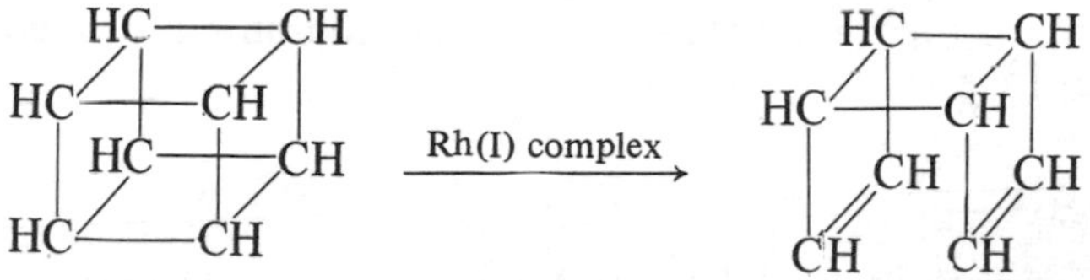

The related '*trans*'-diene is converted by silver fluoroborate into cyclo-octatetraene.

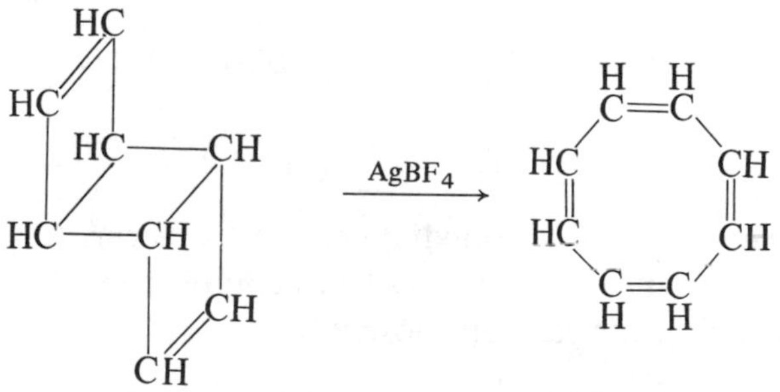

The precise role of the catalyst in these various isomerizations is still a matter of controversy, but mechanisms involving both simple carbonium ions and metal-stabilized carbonium ions (metal–carbene complexes) are commonly invoked (see Sections 2.3 and 2.9). The use of different metal complexes can lead to the formation of different isomeric products. [108, 109] For instance, in contrast to the rhodium-catalysed reaction, cubane is isomerized by silver fluoroborate into cuneane.

$$\begin{array}{c} \text{HC} \underline{\quad} \text{CH} \\ \text{HC} \underline{\quad} \text{CH} \\ \text{HC} \underline{\quad} \text{CH} \\ \text{HC} \underline{\quad} \text{CH} \end{array} \quad \xrightarrow{\text{AgBF}_4} \quad \begin{array}{c} \text{H} \\ \text{C} \\ \text{HC} \underline{\quad} \text{CH} \quad \text{CH} \\ \text{CH} \\ \text{HC} \underline{\quad} \text{CH} \\ \text{CH} \end{array}$$

Strained carbon–carbon σ-bonds are also found in cyclopropanes, and especially the bicyclo[1.1.0]butanes and related structures. Many examples of isomerization of these compounds via organometallic intermediates are known. [110–113]

### 4.2.2  *Alkenes*

Complexes of several Group VIII metals catalyse the migration of double bonds in linear alkenes. [114] For example, 1-octene is isomerized in acetic acid solution in the presence of palladium chloride to yield an equilibrium mixture of 2-octene, 3-octene and 4-octene. No branched chain isomers are formed, and the reaction mechanism is thought to involve a hydrido-π-alkene complex which can then isomerize to a σ-bonded alkyl complex. [115]

$$\begin{array}{c} RCH_2CH{=}CH_2 \\ \downarrow \\ Pd \\ \diagup \mid \diagdown H \end{array} \quad \rightleftharpoons \quad \begin{array}{c} RCH_2CHCH_3 \\ \mid \\ Pd \\ \diagup \mid \diagdown \end{array}$$

The alkyl organometallic intermediate can either revert to the original hydrido complex or can form a new hydrido complex in which the double bond is now located in the adjacent position.

$$\begin{array}{c} RCH_2CHCH_3 \\ \mid \\ Pd \\ \diagup \mid \diagdown \end{array} \quad \rightleftharpoons \quad \begin{array}{c} RCH{=}CHCH_3 \\ \downarrow \\ Pd \\ H \diagup \mid \diagdown \end{array}$$

This new π-complex can then undergo a displacement reaction involving a molecule of unchanged terminal alkene.

$$\begin{array}{c} RCH{=}CHCH_3 \\ \downarrow \\ Pd \\ H \diagup \mid \diagdown \end{array} + RCH_2CH{=}CH_2 \longrightarrow RCH{=}CH{-}CH_3 + \begin{array}{c} RCH_2CH{=}CH_2 \\ \downarrow \\ Pd \\ H \diagup \mid \diagdown \end{array}$$

On this basis *inter*molecular transfer of hydrogen from one molecule to another is involved, and only a trace of hydrido-$\pi$-alkene is required to bring about the isomerization. This latter species is thought to arise from traces of nucleophiles present in the reaction mixture, such as chloride or acetate ions.

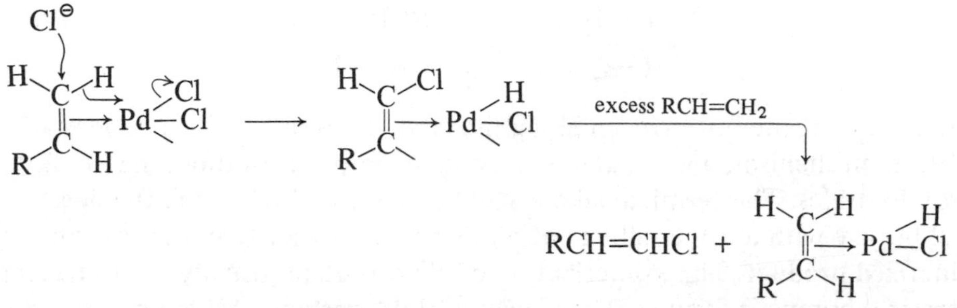

Because of the product distribution this isomerization is not an important synthetic process, but it has mechanistic implications in several reactions involving the synthesis of carbon–hydrogen bonds (Chapter 6.5) and of carbon–oxygen bonds, including an industrial synthesis of ethanal (acetaldehyde) from ethene (see Section 7.6).

A reaction of major synthetic utility is the isomerization of non-terminal into terminal alkenes via intermediate organoboron compounds. This conversion is made possible by the reversible nature of the hydroboration process and the consequent thermal isomerization of alkyl boranes. [116, 117]

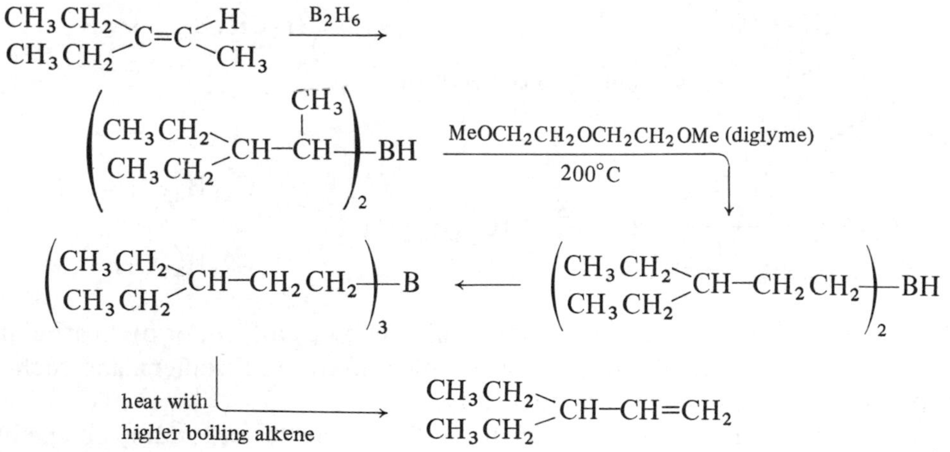

The initial addition of borane to the alkene occurs via a four-centre transition state; the addition is *cis* and is governed primarily by steric interactions so that the boron atom becomes attached to the less substituted of the carbon atoms of the initial double bond.

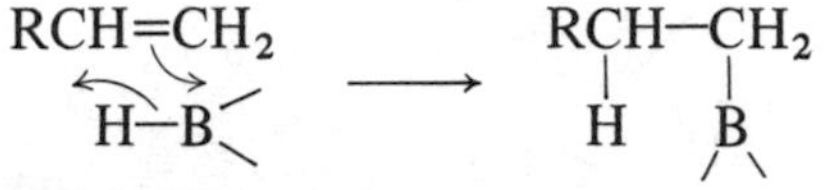

Migration of the boron atom along the chain then occurs by an elimination-addition mechanism, and is catalysed by small amounts of diborane or alkyl boron hydrides. The terminal alkene can be liberated by heating the new alkyl borane with another alkene of boiling point higher than that of the isomerized product. The isomerization of alkyl boranes usually continues until a terminal borane is formed. Thus a new carbon–carbon double bond can be developed at a position quite remote from that in the initial alkene.

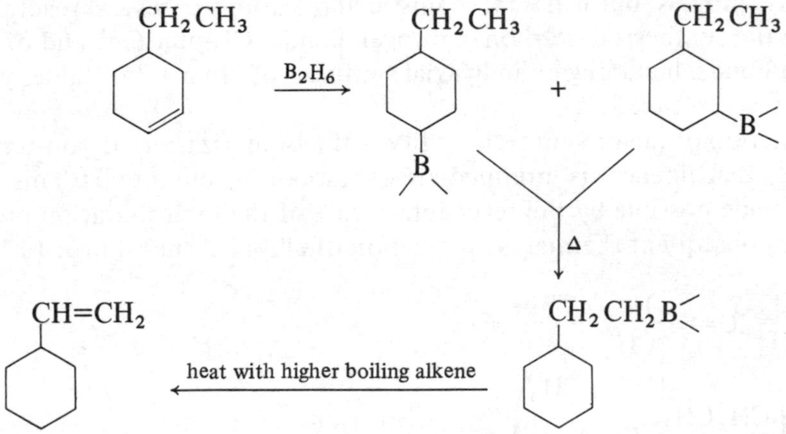

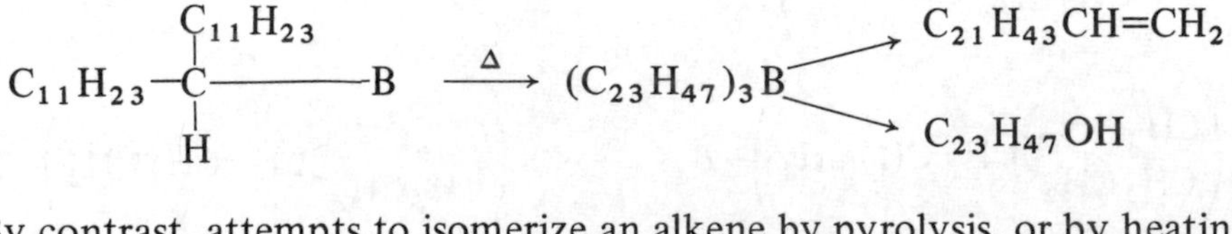

By contrast, attempts to isomerize an alkene by pyrolysis, or by heating it with various acids, usually result in a complex mixture of isomers, and such processes have very little synthetic utility.

The alkyl borane can also be transformed into the corresponding alkane by treatment with acetic acid, or into the corresponding alcohol by treatment with alkaline hydrogen peroxide (see Sections 6.3 and 7.3 respectively).

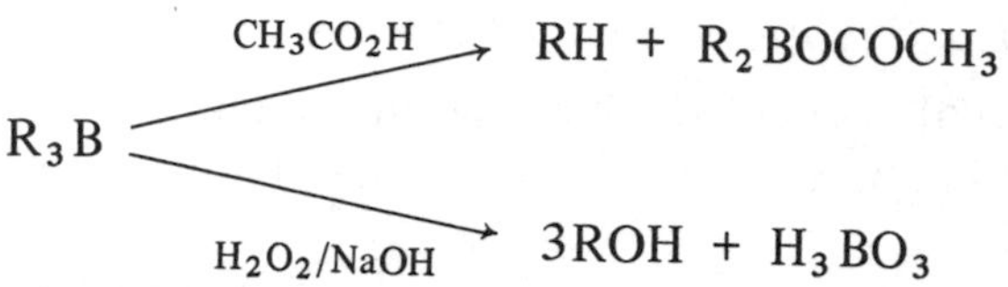

$$R_3B \begin{cases} \xrightarrow{CH_3CO_2H} RH + R_2BOCOCH_3 \\ \xrightarrow{H_2O_2/NaOH} 3ROH + H_3BO_3 \end{cases}$$

### 4.2.3  *Alkadienes*

Metal $\pi$-complexes are implicated as intermediates in the isomerization of dienes, and the usual result is the conversion of the unconjugated into its conjugated isomer. However, this generalization does not always hold, as exemplified by the reverse isomerization of 1,3-cyclo-octadiene to the 1,5-isomer by rhodium(III) trichloride. [118]

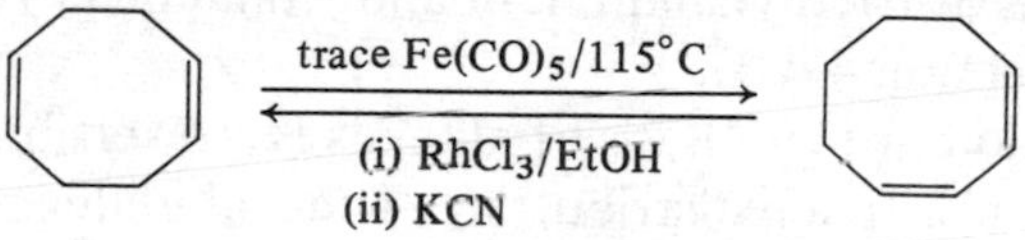

trace Fe(CO)$_5$/115°C
(i) RhCl$_3$/EtOH
(ii) KCN

At present, these isomerization processes are not fully understood and it is difficult to predict the outcome of a particular reaction (see however Section 4.2.2 and Chapter 6.5). This is an area of active research, and synthetic use of these isomerizations will undoubtedly increase with a further understanding of the underlying principles.

One reaction of this type which is widely used is the isomerization of a 1,4-diene to a 1,3-diene by tris(triphenylphosphine)chlororhodium in refluxing chloroform solution. [119]

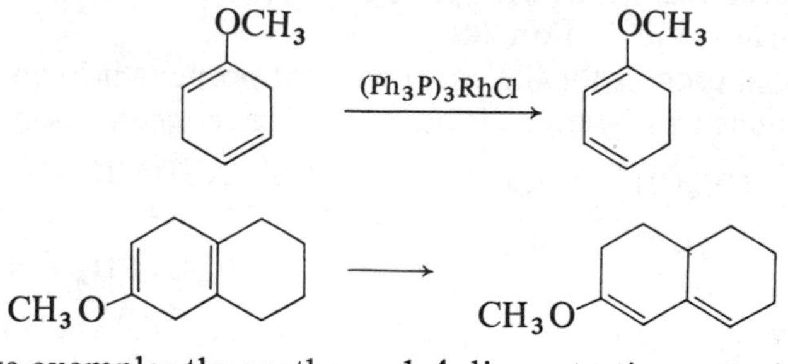

(Ph$_3$P)$_3$RhCl

In the above examples the methoxy 1,4-diene starting materials are readily available from the metal–ammonia reduction of the related methyl aryl ethers.

69

## 4.3 Dimerization

### 4.3.1 *Dimerization and reorganization of alkenes*

Simple dimerizations of alkenes which are neither oxidative nor reductive are frequently observed in photochemical reactions, but are unusual in the organometallic field.

The dimerization of ethene to a mixture of *cis*- and *trans*-2-butene is catalysed by rhodium(III) trichloride, but the process is not general for higher alkenes. [120]

$$2 \; \underset{H}{\overset{H}{\phantom{.}}}C=C\underset{H}{\overset{H}{\phantom{.}}} \xrightarrow[45°C]{RhCl_3} \quad \underset{H}{\overset{CH_3}{\phantom{.}}}C=C\underset{H}{\overset{CH_3}{\phantom{.}}} + \underset{H}{\overset{CH_3}{\phantom{.}}}C=C\underset{CH_3}{\overset{H}{\phantom{.}}}$$

The catalyst system $Ti(OC_4H_9)_4/AlEt_3$ dimerizes ethene to 1-butene, but most combinations of triethylaluminium and titanium(IV) salts lead to polymerizations (see Section 4.4.4).

The reorganization or metathesis of alkenes is a reversible reaction involving bond reorganization with a redistribution of the alkylidene moieties. [121, 122]

$$\begin{array}{c} RCH=CHR' \\ + \\ RCH=CHR' \end{array} \xrightleftharpoons{\text{metal catalyst}} \begin{array}{cc} RCH & HCR' \\ \| & + \; \| \\ RCH & HCR' \end{array}$$

The heterogeneous catalyst initially used in this reaction was based on $MoO_3 \cdot Al_2O_3$. Homogeneous catalysts derived from $MoCl_5$ or $WCl_6$ and organolithium or organoaluminium compounds have since been reported. The active catalytic species involves a lower-valence form of the transition metal, and reactions proceed readily at room temperature in benzene solution. A possible mechanism is shown in Section 2.9.

A commercial process for the conversion of propene into an equimolar mixture of ethene and 2-butene is based on this reorganization reaction.

$$\begin{array}{cc} CH_3CH & HCCH_3 \\ \| & + \; \| \\ HCH & HCH \end{array} \rightleftharpoons \begin{array}{c} CH_3CH=CHCH_3 \\ + \\ H_2C=CH_2 \end{array}$$

The bond reorganization of cycloalkenes has a special application in the preparation of macrocyclic dienes in good yield under high dilution conditions. [123]

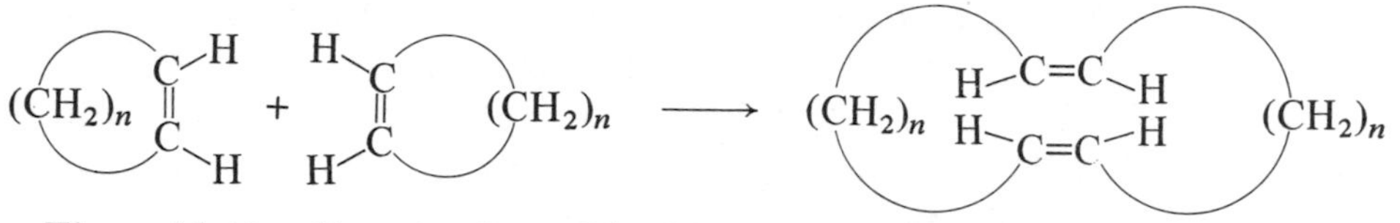

The oxidative dimerization of β-alkyl and -aryl substituted terminal alkenes can be achieved with a combination of palladium acetate and sodium acetate in acetic acid. [124]

$$\begin{matrix}CH_3\\C_6H_5\end{matrix}\!>\!C{=}CH_2 \xrightarrow[\text{HOAc}]{Pd(OAc)_2/NaOAc} \begin{matrix}CH_3\\C_6H_5\end{matrix}\!>\!C{=}CH{-}CH{=}C\!<\!\begin{matrix}CH_3\\C_6H_5\end{matrix}$$

Deuterium labelling studies show that hydrogen abstraction takes place exclusively from the terminal vinylic carbon atom of the co-ordinated alkene with no loss of hydrogen from the methyl group. Thus, the formation of an intermediate π-allyl palladium complex is excluded and a binuclear π-alkene palladium acetate complex is postulated. Oxidation of the alkene is accompanied by reduction of the palladium salt to palladium metal and the formation of two moles of acetic acid.

$$2\ {>}C{=}CH_2 + Pd(OAc)_2 \longrightarrow {>}C{=}CH{-}CH{=}C{<} + Pd + 2HOAc$$

Similar oxidative dimerization reactions have been reported for various arenes. [125]

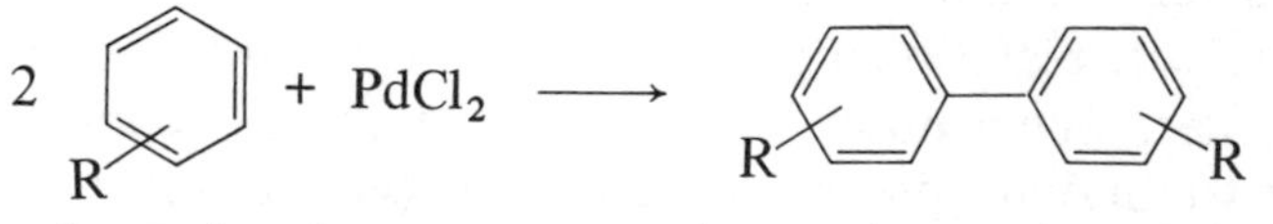

The action of powdered manganese and organic or mineral acids causes dimerization of dialkyl maleates to yield only the enantiomeric butane-1,2,3,4-tetracarboxylates, possibly via an organo–manganese intermediate. [126] Dialkyl fumarates fail to react.

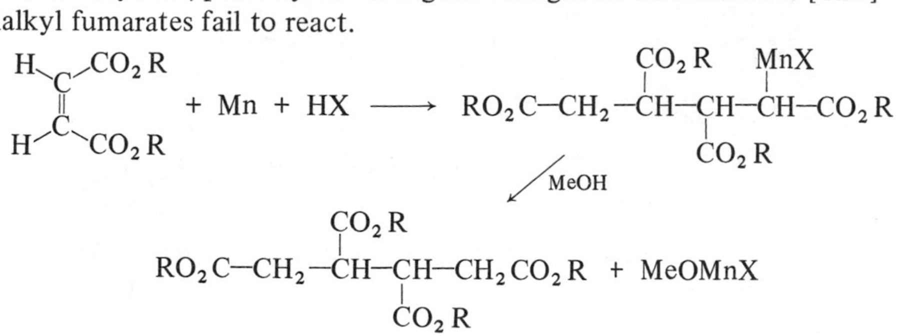

Methods involving organometallic intermediates are known for the head-to-head reductive dimerization of alkenes. The overall process can be shown $RCH=CH_2 \rightarrow RCH_2CH_2CH_2CH_2R$. When a terminal alkene is heated with metallic sodium in diethyl ether, reductive dimerization and polymerization can occur, and also reduction to the corresponding monomeric hydrocarbon. In the presence of finely divided sodium, an 'electron carrier' such as terphenyl, and a complexing ether such as tetrahydrofuran, formation of the reductive dimer is favoured. A radical anion, $(R\overset{\ominus}{C}H-CH_2)Na^{\oplus}$ is postulated as the intermediate, and the hydrocarbon product is formed by protonation during work up.

As discussed in Section 4.2.2, alkenes may be converted into trialkylboranes, which on oxidation with silver nitrate in the presence of sodium hydroxide yield the dimeric hydrocarbons. [127]

$$2(RCH_2CH_2)_3B \xrightarrow{\text{AgNO}_3/\text{NaOH}} 3RCH_2CH_2CH_2CH_2R$$

Yields are highest in the case of primary alkylboranes, and the mechanism is thought to involve free radicals. Yields in attempted cross couplings using two different alkylboranes are poor.

### 4.3.2 *Dimerization of alkadienes*

The direct dimerization of alkadienes has been observed only in special cases. 1,3-Butadiene with bis(triphenylphosphite)dicarbonylnickel yields *cis,cis*-1,5-cyclo-octadiene which is now an important industrial chemical. On the other hand, with a triethylphosphine aryl nickel complex, butadiene undergoes cyclo-dimerization to yield 2-methylenevinylcyclopentane, and with various other catalysts a wider range of dimers can be obtained. [128, 129]

Dimeric hydrocarbons are also derived from norbornadiene by reaction with nonacarbonyldiiron or other more complex organometallic compounds. [130, 131]

72

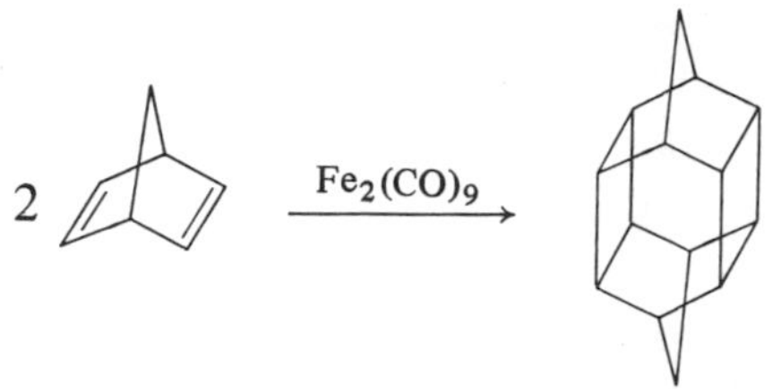

$$2 \quad \xrightarrow{Fe_2(CO)_9}$$

Numerous palladium catalysed dimerizations of alkadienes involving incorporation of other molecules have been reported. [132] One example is the dimerization of butadiene in the presence of an aldehyde to yield a 2-substituted-3,6-divinyltetrahydropyran.

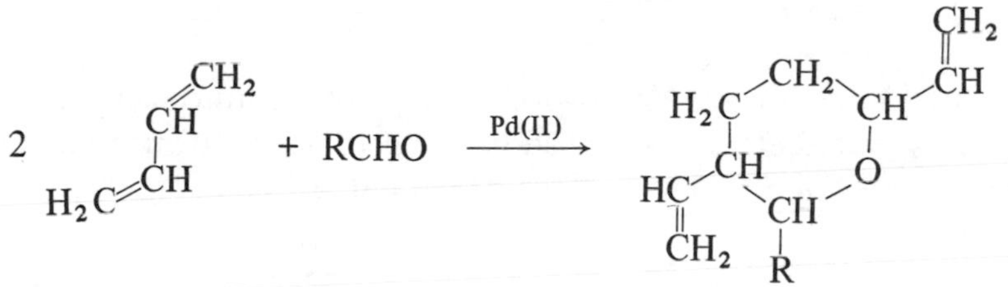

Naphthalene-lithium, $(C_{10}H_8^{\ominus})Li^{\oplus}$, converts 1,3-cyclo-octadiene into the reductive dimer shown below; it also dimerizes isoprene (2-methyl-1,3-butadiene) to a mixture of 2,6-dimethyl-2,6-octadiene (head to tail linkage) and 2,7-dimethyl-2,6-octadiene (head to head linkage). Radical anions are again likely intermediates, and protonation can occur either in the reaction mixture or during work-up. [133, 134]

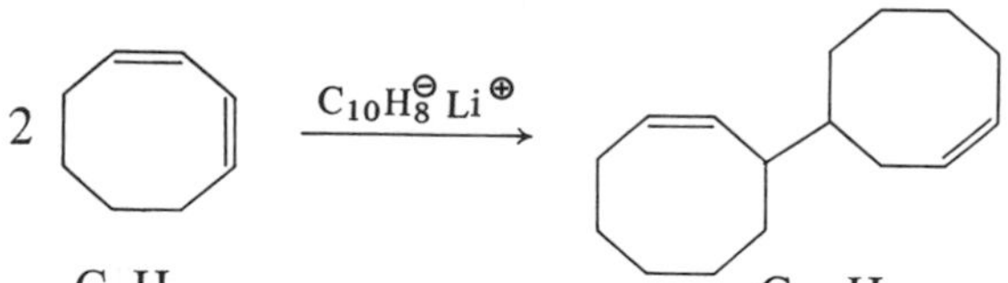

$C_8H_{12}$ $\qquad$ $C_{16}H_{26}$

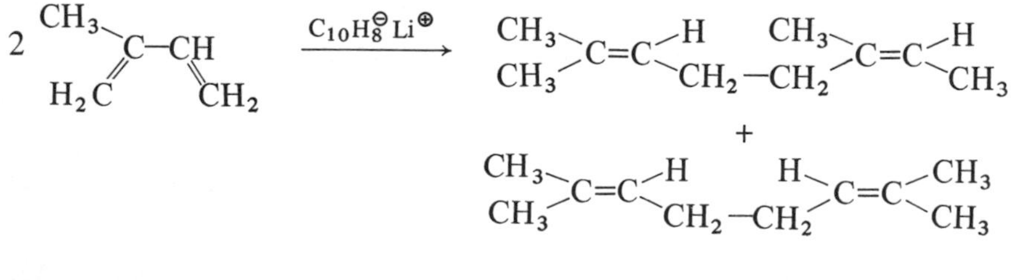

The above reactions are closely related to the phenomenon of 'reduction by dissolving metals' of which the Birch reduction of aromatic ethers by sodium in liquid ammonia is an important example.

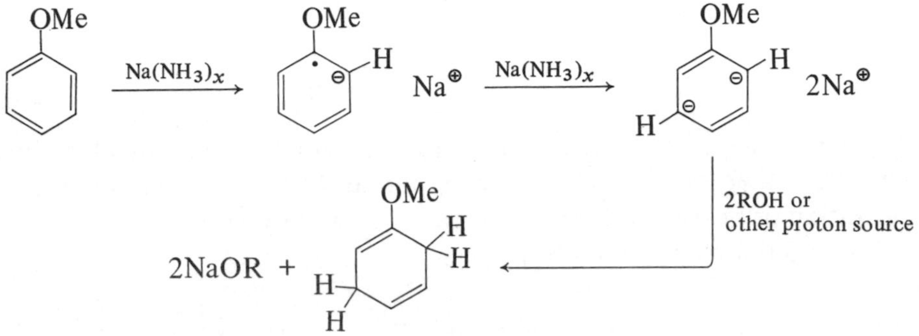

The blue colour of a solution of sodium in liquid ammonia is due to the presence of solvated electrons; the addition of these electrons to an organic substrate is equivalent to an electrochemical reduction.

$$Na \rightleftharpoons Na^{\oplus} + e(NH_3)_x^{\ominus}$$

The addition of one electron to a neutral organic molecule produces a radical anion; in general this can react either with another electron to produce a dianion or else can undergo protonation *in situ* to give a radical which can then undergo normal radical reactions, including dimerization:

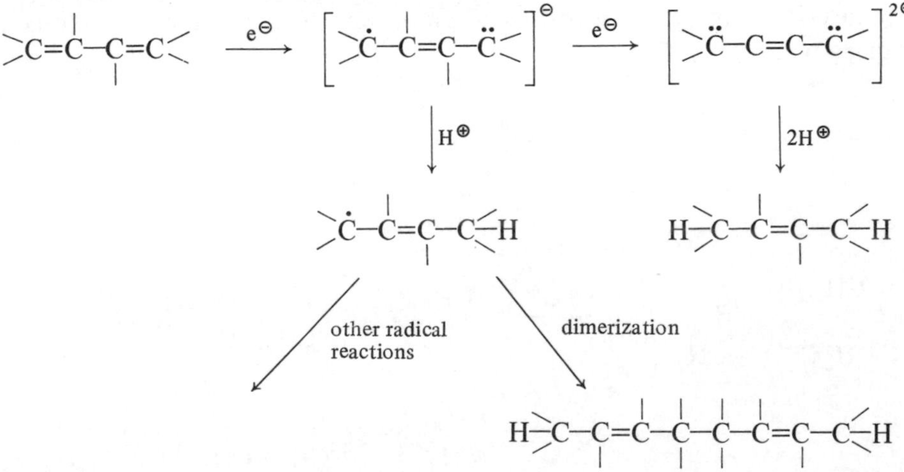

A solution of naphthalene-lithium in tetrahydrofuran or of sodium or potassium in ethylenediamine acts as a source of electrons in the same way as a solution of sodium in liquid ammonia.

### 4.3.3  *Dimerization of alkynes*

The dimerization of a terminal alkyne to yield an enyne (vinyl acetylene) can be effected by dissolving a catalytic amount of copper(I) oxide in hot acetic acid and adding the alkyne. [16, 135] A copper(I) acetylide is probably an intermediate.

$$2RC{\equiv}CH \xrightarrow{\text{Cu}_2(\text{OAc})_2} RCH{=}CH{-}C{\equiv}CR$$

Most other alkyne dimerizations are either oxidative or reductive. Anhydrous copper(II) acetate in methanol-pyridine brings about the oxidative dimerization of terminal alkynes to di-ynes, a free radical reaction already discussed in Section 2.4. The species which undergoes oxidation is a copper(I) acetylide.

$$2RC{\equiv}CH \xrightarrow{\text{Cu}(\text{OAc})_2} RC{\equiv}C{-}C{\equiv}CR$$

Reductive dimerization of terminal alkynes has been achieved by reduction of the alkyne with diisobutylhydridoaluminium to yield a vinyl aluminium compound (cf. Section 3.5) which is then dimerized by the addition of copper(I) chloride. [136] The latter step is again presumably of free radical character (cf. the free radical coupling of vinyl magnesium compounds, Sections 2.4 and 3.4).

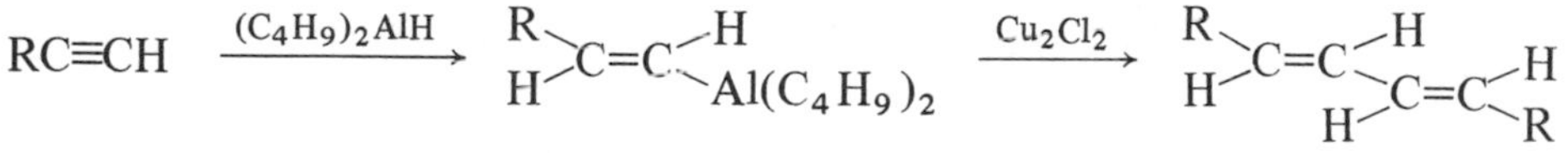

Direct cyclodimerization of alkynes can be effected by a variety of transition metal complexes such as cyclopentadienyldicarbonylcobalt and bis-(benzonitrile)dichloropalladium. In each case the expected product, a cyclobutadiene, is obtained only as its metal complex.

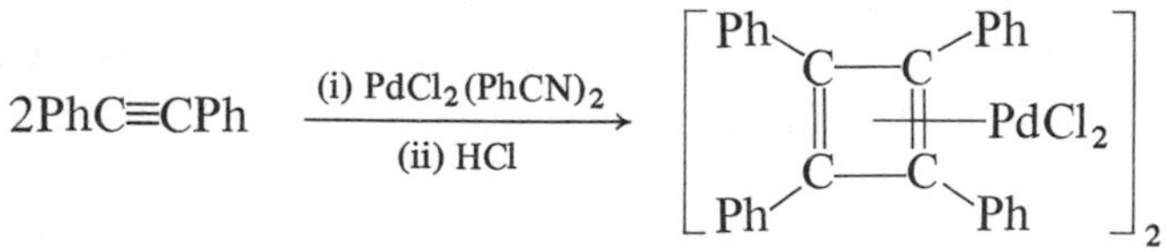

75

Using other catalysts or other reaction conditions, trimerization and/or tetramerization of the alkyne substrate leads to derivatives of benzene or of cyclo-octatetraene (see Section 4.4.2 below).

A strained cyclopropene behaves in many respects like an alkyne, and 1-methylcyclopropene can be dimerized by palladium chloride to a mixture of two tricyclic isomers. The function of the metal in this reaction is not clear, but presumably the first intermediate is a palladium $\pi$-complex. [137]

$$2CH_3-C{=}CH \xrightarrow[]{\ \ PdCl_2\ \ } \begin{array}{c} CH_2 \\ CH_3-C-CH \\ | \quad | \\ CH_3-C-CH \\ CH_2 \end{array} + \begin{array}{c} CH_2 \\ CH_3-C-CH \\ | \quad | \\ HC-C-CH_3 \\ CH_2 \end{array}$$

## 4.4 Higher polymerization

### 4.4.1 *Cyclo-oligomerization of alkadienes*

The cyclotrimerization of 1,3-butadiene can be carried out using the combination of a suitable transition metal compound with an aluminium alkyl. Triethylaluminium/chromium(III) chloride gives mainly the *trans, trans, trans*-1,5,9-cyclododecatriene, while diethylchloroaluminium/titanium tetrachloride gives mainly the corresponding *cis, trans, trans* isomer (cf. Section 2.1).

$$3CH_2{=}CH{-}CH{=}CH_2 \quad \begin{array}{c} \xrightarrow[]{\ Et_3Al/CrCl_3\ } \\ \\ \xrightarrow[]{\ Et_2AlCl/TiCl_4\ } \end{array}$$

The nickel complex of *trans, trans, trans*-1,5,9-cyclododecatriene is itself able to catalyse trimerization of 1,3-butadiene to the same cyclic triene by simultaneously forming a new but identical complex and releasing 1,5,9-cyclododecatriene from the initial complex.

It is suggested that three molecules of butadiene join together on the nickel complex as template to give an intermediate species in which two molecules of the cyclotriene are linked to the nickel atom. This complex dissociates to regenerate a molecule of product and one of the initial catalyst which then continues the process.

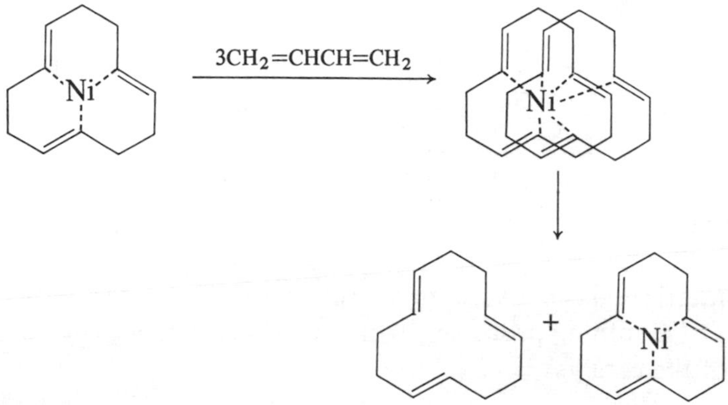

$$3CH_2=CHCH=CH_2$$

Allene (1,2-propadiene) also undergoes cyclo-oligomerization to a mixture of products in the presence of bis(triphenylphosphine)dicarbonylnickel and related complexes.

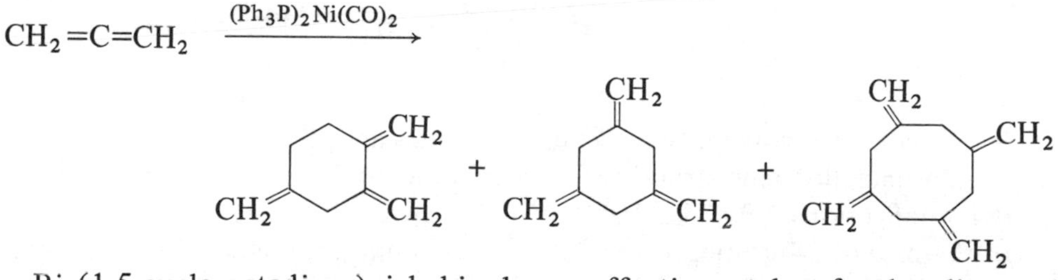

$$CH_2=C=CH_2 \xrightarrow{(Ph_3P)_2Ni(CO)_2}$$

Bis(1,5-cyclo-octadiene)nickel is also an effective catalyst for the oligomerization of dienes. [128]

## 4.4.2 Cyclo-oligomerization of alkynes

It was discovered as early as 1940 that cyclo-octatetraene could be produced by the reaction of acetylene with nickel(II) cyanide in tetrahydrofuran, and that benzene was a by-product.

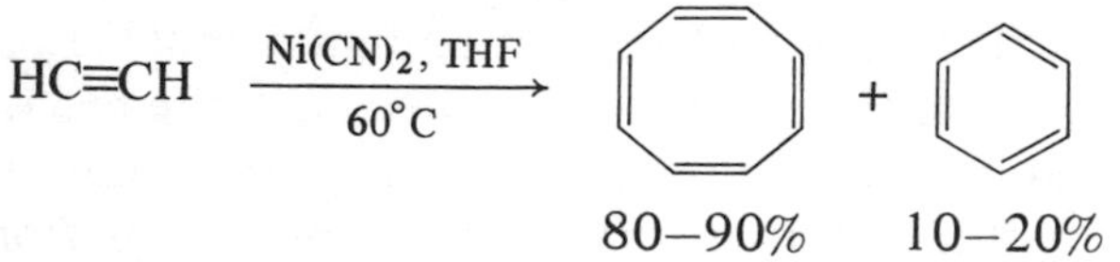

$$HC\equiv CH \xrightarrow[60°C]{Ni(CN)_2, THF}$$

80–90%      10–20%

77

Since that time, many metal complexes including those of nickel, cobalt, palladium, rhodium and iron, and also Ziegler–Natta catalysts (see Section 4.4.3), have been used as catalysts in the formation of arynes from substituted alkynes.

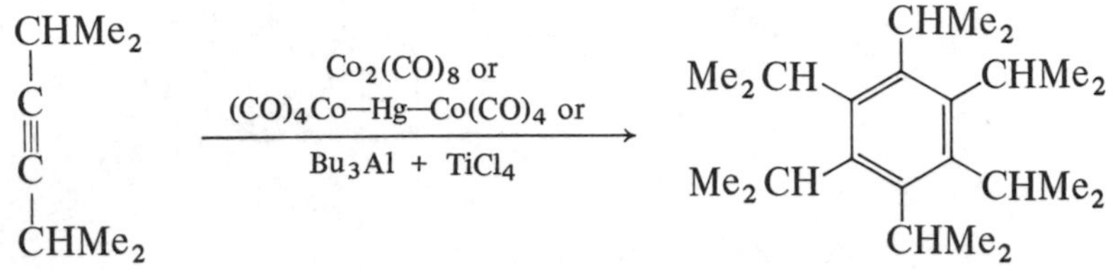

Unsymmetrical alkynes yield mixtures of arenes as a result of head-to-head and head-to-tail joining processes, which can be influenced to some extent by selection of the catalyst.

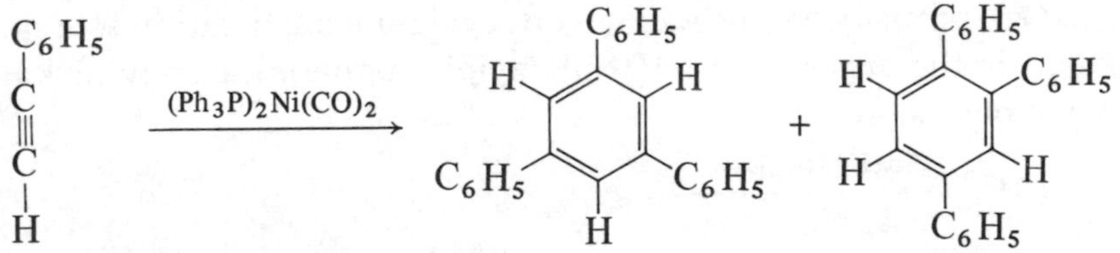

Intermediates involving two alkyne molecules are postulated: these can be written as metallocyclic structures or as metal-complexed cyclobutadienes [138] (cf. Sections 2.9 and 2.10).

In the case of 2-butyne, trimerization with aluminium chloride in benzene leads to 'hexamethyl Dewar benzene'. [139]

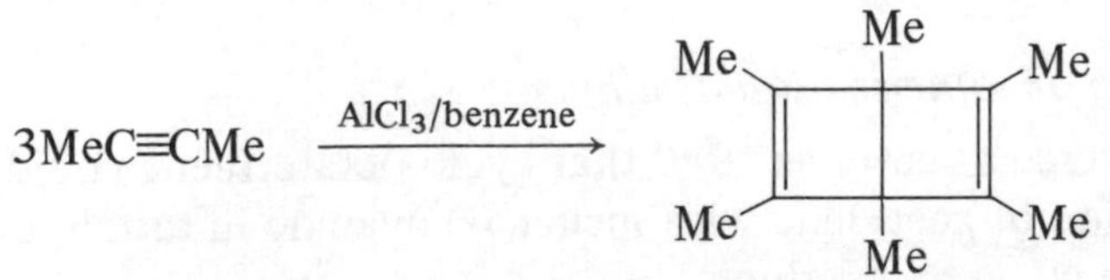

In some reactions of alkynes with aryl chromium complexes, an aryl group is incorporated into the organic product. For example, the reaction of triphenylchromium and 2-butyne in tetrahydrofuran affords not only hexamethyl benzene, but 1,2,3,4-tetramethylnaphthalene, arising from two molecules of butyne and one phenyl group initially $\sigma$-bonded to the chromium. Under certain conditions the naphthalene compound can be made the major product. [140]

The special virtue of these reactions in synthesis is that they provide routes to tri- or hexa-substituted benzenes and to condensed aromatic hydrocarbons which may be difficult to obtain by other methods.

### 4.4.3  *Polymerization of alkenes*

Many of the dimerization and oligomerization reactions discussed above can also be carried out in the absence of a metal catalyst but with the application of heat and pressure. In such cases, while the products may be the same, the reaction mechanisms are undoubtedly very different.

This is also true of the important alkene polymerizations now to be discussed. Ethene can be converted into a high molecular weight linear polymer by heating the gas under high pressure, and free radical intermediates are clearly involved. By contrast, many alternative metal-catalysed processes are now available for the polymerization of ethene to a variety of linear polymers, with molecular weights ranging from around 150 to several million. Such reactions can be carried out at ordinary pressures and often at low temperatures, and the catalyst systems are such that organometallic intermediates are involved throughout the polymerization. These processes came about through the discoveries of Ziegler in Germany and Natta in Italy; a typical Ziegler–Natta catalyst is the complex insoluble solid precipitated from hydrocarbon solvent by the interaction of triethylaluminium with titanium tetrachloride. Polymerization of alkenes occurs on the surface of such catalysts by ionic mechanisms involving both $\pi$-bonded and $\sigma$-bonded organometallic intermediates.

Reaction on a Ziegler–Natta catalyst surface occurs in a regular and stereo-chemically selective manner, the polymer chain containing uniformly oriented

79

monomer units. A segment of polypropylene chain produced by the Ziegler–Natta polymerization of propene can be shown as:

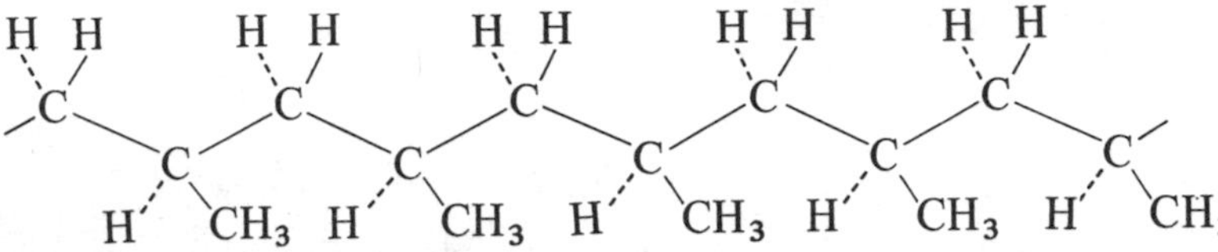

Polypropylene prepared in this way is a hard, strong, high-melting solid said to be 'crystalline' in character. The 'isotactic' hydrocarbon chain adopts a helical conformation in the solid with the methyl groups arranged so as to minimize steric interactions between them. Polyethylene prepared by Ziegler–Natta methods is superior in physical properties to the polymer produced by the high pressure free-radical process, but because of the inherent regularity of the $(CH_2)$ chain, the differences are less pronounced than in the case of polypropylene.

The mechanism of the Ziegler–Natta polymerization process is not fully understood, but there is strong evidence to support the postulate that the aluminium alkyl acts as a reducing and alkylating agent, producing alkylated transition metal atoms of lower valence in the catalyst surface which become the active sites for polymerization. In the co-polymerization of propene and ethene, the relative reactivity of propene is increased as the transition metal is changed from hafnium through zirconium and titanium to vanadium. In contrast, there is no change in the relative reactivities of propene and ethene when organometallic reducing agents other than triethylaluminium are used in combination with the same transition metal complex, giving strong support to the idea that polymerization involves the transition metal atom and not an aluminium atom.

A reasonable mechanism for polymerization is based on initial $\pi$-co-ordination of the monomeric alkene to the alkylated transition metal atom, followed by migration of the attached alkyl group from metal to carbon. Co-ordination of the alkene to the metal atom polarizes the carbon–carbon bond and allows ready migration of the alkyl group with its bonding electron pair (cf. Section 2.6). This alkyl group migration is thought to occur as a concerted process and the $\pi$-bonded alkene is thereby changed to a $\sigma$-bonded alkyl group. The resulting intermediate can then co-ordinate with another alkene molecule. The process continues in this fashion and polymerization occurs with an alkyl

80

chain growing steadily outwards from the co-ordinated metal atom. In the formulae below, $L_n$ represents $n$ co-ordinated ligands, L, and R represents an alkyl group derived from the trialkyl aluminium reducing agent.

$$L_n MR + CH_2{=}CH_2 \longrightarrow L_n M{\leftarrow}\overset{CH_2}{\underset{\underset{R}{|}}{\|}}_{CH_2} \longleftrightarrow L_n \overset{\ominus}{M}{-}CH_2$$

$$\xleftarrow{\text{etc.}} L_n M{\leftarrow}\overset{CH_2}{\underset{\underset{\underset{\underset{R}{|}}{CH_2}}{|}}{\|}}_{CH_2} \xleftarrow{CH_2=CH_2} L_n M{-}CH_2 CH_2 R$$

Ziegler–Natta catalysts are often unstable, the active sites may differ from system to system, and product molecular weights increase with increasing temperature and decreasing solvent polarity. However, the above polymerization mechanism involving co-ordination of the alkene to a transition metal complex followed by migration of the growing alkyl chain from metal atom to a carbon atom of the co-ordinated alkene is probably quite general.

Alkenes such as ethene and propene can also be polymerized to lower molecular weight products using triethylaluminium alone. The essential mechanistic feature of this reaction is again $\pi$-co-ordination of the alkene to a metal atom, in this case aluminium, followed by migration of an alkyl group from aluminium to an alkene carbon atom.

$$\underset{Et}{\overset{Et}{\diagdown}}Al{-}Et \xrightarrow{CH_2=CH_2} \overset{CH_2=CH_2}{\underset{\underset{Et}{|}}{\underset{Et}{\diagup}}Al}\diagdown_{Et} \longleftrightarrow \overset{\overset{\oplus}{CH_2}CH_2}{\underset{\underset{Et}{|}}{Et{\diagup}\overset{\ominus}{Al}}\diagdown_{Et}}$$

$$\xleftarrow{\text{etc.}} \xleftarrow{CH_2=CH_2} \overset{CH_2 CH_2 Et}{\underset{\underset{Et\diagup \diagdown Et}{|}}{Al}}$$

The process can then continue with insertion of further alkene units; the final reaction product of triethylaluminium and ethene can be written:

$$Et(CH_2CH_2)_l-Al \underset{(CH_2CH_2)_n Et}{\overset{(CH_2CH_2)_m Et}{<}}$$

where $l, m$, and $n$ are integers.

The reaction can be controlled to give products having molecular weights within a fairly narrow range; subsequent treatment of the trialkylaluminium with acid yields the corresponding hydrocarbons. More importantly, oxidation followed by hydrolysis yields long-chain primary alcohols (see Section 7.2.3), while thermal decomposition yields terminal alkenes and aluminium hydride (cf. Section 4.1.1).

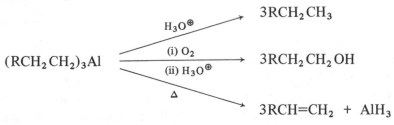

$$(RCH_2CH_2)_3Al \nearrow \overset{H_3O^\oplus}{\longrightarrow} 3RCH_2CH_3$$

$$\overset{(i)\ O_2}{\underset{(ii)\ H_3O^\oplus}{\longrightarrow}} 3RCH_2CH_2OH$$

$$\overset{\Delta}{\searrow} 3RCH=CH_2 + AlH_3$$

Simple lithium alkyls also undergo addition to alkenes which are consequently polymerized.

$$LiC_3H_7 \overset{CH_2=CH_2}{\longrightarrow} LiCH_2CH_2C_3H_7 \overset{CH_2=CH_2}{\longrightarrow} Li(CH_2CH_2)_2C_3H_7 \overset{etc.}{\longrightarrow}$$

It was this early discovery by Ziegler which led to the important industrial processes for alkene polymerization discussed above. Industrial application was greatly facilitated by the discovery of a cheap two-step process for the large-scale synthesis of triethylaluminium from ethene, hydrogen, and aluminium metal:

$$2Et_3Al + Al + \tfrac{3}{2}H_2 \overset{150°C}{\longrightarrow} 3Et_2AlH$$

$$3Et_2AlH + 3CH_2=CH_2 \overset{70°C}{\longrightarrow} 3Et_3Al$$

$$\text{Sum:}\quad Al + \tfrac{3}{2}H_2 + 3CH_2=CH_2 \longrightarrow Et_3Al$$

Further research in this important field continues actively at the present time.

# Carbon monoxide insertion reactions

5

## 5.1 Introduction

As pointed out in Section 2.6, carbon monoxide insertion reactions are best regarded as belonging to the class of group migration reactions in which the migrating group moves from a metal atom to a carbon monoxide moiety attached to the same metal atom. A new carbon–carbon bond is thus formed, the carbonyl carbon remaining attached to the metal. Subsequent reaction of the organometallic intermediate can lead to a variety of products such as aldehydes, ketones, alcohols, acids, or esters, according to the conditions used or other reagents present. Transition metals such as iron, cobalt, nickel, rhodium and palladium are principally involved in these diverse carbonyl insertion reactions. Recent developments by Brown have opened up many new possibilities in reaction of organoboranes with carbon monoxide, and in these cases a succession of group migrations from the boron atom to the carbon atom of the original carbon monoxide molecule is possible.

The carbon monoxide insertion reactions of transition metal complexes will be presented first, since these are closely related to the alkene polymerizations discussed above in Section 4.4.3.

## 5.2 Hydroformylation

The industrially important 'oxo' or hydroformylation reaction, which involves the addition of carbon monoxide and hydrogen to an alkene catalysed by tricarbonylhydridocobalt, has been discussed briefly in Section 2.6. The product is an aldehyde having one more carbon than the initial alkene.

$$RCH{=}CH_2 \xrightarrow[\text{[HCo(CO)}_3]}{\text{H}_2/\text{CO}} RCH_2CH_2CHO$$

Tetracarbonylhydridocobalt can be prepared in toluene or hexane solution from octacarbonyldicobalt and acid in the presence of some dimethylformamide; it is also formed rapidly from octacarbonyldicobalt and various other cobalt compounds under the conditions of the hydroformylation reaction.

The first step of the catalytic reaction is dissociation of a carbonyl group from $HCo(CO)_4$ and its replacement in an equilibrium reaction with a co-ordinated molecule of the alkene. [141–143]

$$HCo(CO)_4 \ + \ RCH{=}CH_2 \ \rightleftharpoons \ HCo \overset{\underset{\displaystyle CO}{|}}{\underset{\underset{\displaystyle CO}{|}}{\overset{\displaystyle CO}{}}} \begin{smallmatrix} CHR \\ \| \\ CH_2 \end{smallmatrix} \ + \ CO$$

Migration of a hydride ion from cobalt to carbon then gives rise to an alkyltricarbonyl cobalt; both isomers are formed but the linear compound usually predominates.

$$H{-}Co \overset{CO}{\underset{CO}{\lessgtr}} \ \ \longrightarrow \ \ RCH_2CH_2{-}Co{-}CO \ + \ CH_3CH{-}Co{-}CO$$

This type of addition of a metal hydride to an alkene is well known for boron, aluminium and other metal hydrides (cf. Sections 3.5, 3.6, 4.2.2, 4.3.3, 6.5, and 7.3.1).

A further equilibrium is then established between this alkyltricarbonylcobalt and the corresponding tetracarbonyl compound; the latter is unstable and migration of the alkyl group from cobalt to one of the carbonyl carbon atoms occurs. [144, 145]

$$RCH_2CH_2{-}Co{-}CO \ \underset{-CO}{\overset{+CO}{\rightleftharpoons}} \ RCH_2CH_2{-}Co{\lessgtr}^{CO}_{CO} \ \longrightarrow \ RCH_2 \begin{smallmatrix} Co \\ | \\ CH_2 \end{smallmatrix} C{=}O$$

The acyltricarbonylcobalt product undergoes hydrogenolysis via an intermediate dihydrido complex to give the product aldehyde with regeneration of the tricarbonylhydridocobalt catalyst, which can then co-ordinate with a fresh molecule of alkene. [146, 147]

84

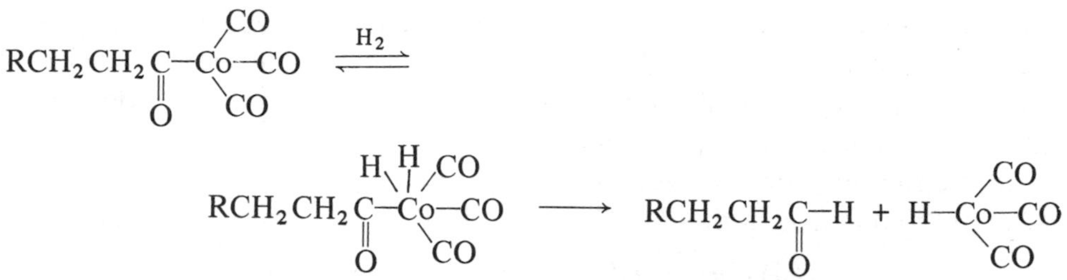

$$RCH_2CH_2\underset{\underset{O}{\parallel}}{C}\!-\!\underset{\underset{CO}{\diagdown}}{\overset{\overset{CO}{\diagup}}{Co}}\!-\!CO \;\overset{H_2}{\underset{\rightleftharpoons}{}}$$

$$RCH_2CH_2\underset{\underset{O}{\parallel}}{C}\!-\!\underset{\underset{CO}{\diagdown}}{\overset{\overset{H\;\;H\;\;CO}{|\;/}}{Co}}\!-\!CO \;\longrightarrow\; RCH_2CH_2\underset{\underset{O}{\parallel}}{C}\!-\!H \;+\; H\!-\!\underset{\underset{CO}{\diagdown}}{\overset{\overset{CO}{\diagup}}{Co}}\!-\!CO$$

This whole process can be summarized as shown below; each successive metal complex has alternatively 16 or 18 valence electrons associated with the cobalt atom (cf. Chapter 2.5).

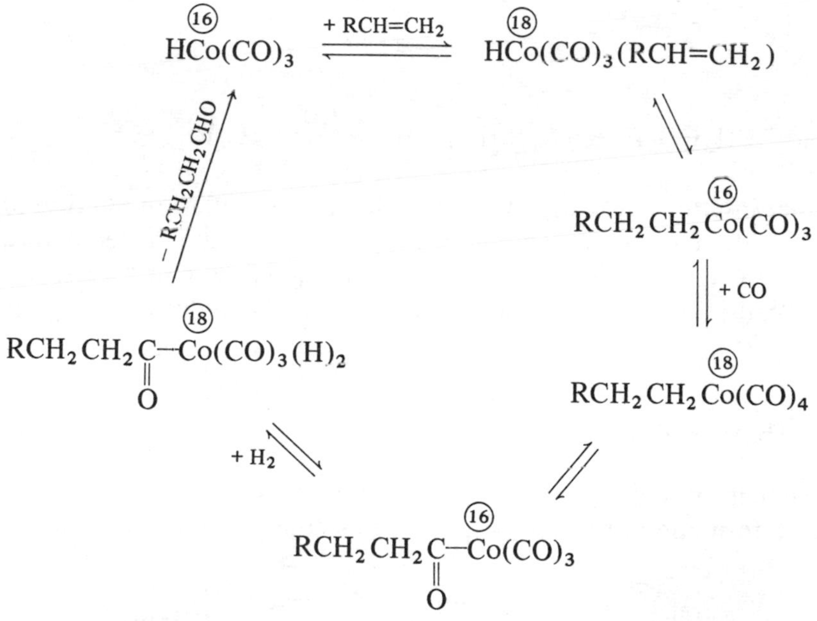

Rates of hydroformylation reactions decrease in the order linear terminal alkenes, linear non-terminal alkenes, and more highly substituted alkenes, presumably reflecting increasing difficulty of the initial metal hydride addition reaction. If the hydrogenation is allowed to continue after all the carbon monoxide is used up, the product is an alcohol rather than an aldehyde.

Various rhodium catalysts are active in the hydroformylation of alkenes and alkynes. [148] Homogeneous rhodium catalysts can also activate carbon monoxide at low pressures and allow the conversion of methyl iodide into acetic acid. This is the basis of an industrial process for the preparation of acetic acid from methanol and the proposed mechanism involves oxidative addition of methyl iodide to a rhodium(I) atom, migration of a methyl group from the rhodium atom to a carbonyl carbon atom, and reductive elimination of acetyl iodide from a rhodium(III) atom [149] (see Sections 2.5 and 2.6).

$$CH_3OH + HI \longrightarrow CH_3I + H_2O$$

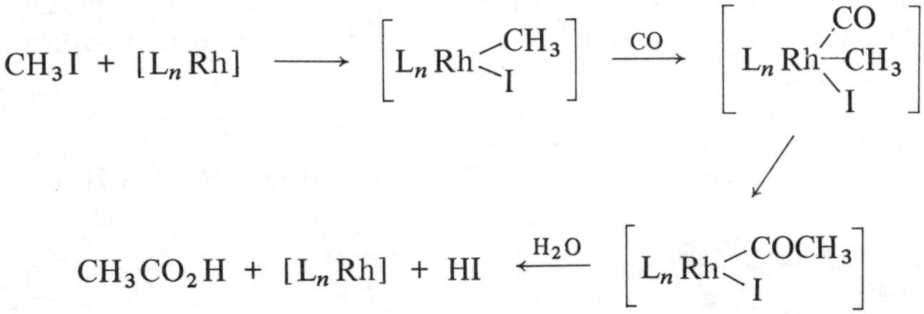

$$CH_3I + [L_nRh] \longrightarrow \left[ L_nRh{<}^{CH_3}_{I} \right] \xrightarrow{CO} \left[ L_nRh{<}^{CO}_{CH_3} \atop I \right]$$

$$CH_3CO_2H + [L_nRh] + HI \xleftarrow{H_2O} \left[ L_nRh{<}^{COCH_3}_{I} \right]$$

Similar intermediates are involved in the decarbonylation of aroyl halides to aryl halides by tris(triphenylphosphine)chlororhodium at temperatures above 200°C (see also Section 7.8). In refluxing benzene or toluene this same catalyst causes the decarbonylation of aliphatic aldehydes to alkanes in good yield. [150]

## 5.3    Hydrocarboxylation

Replacement of hydrogen in the cobalt-catalysed oxo reaction by an alcohol results in formation of the corresponding ester as the major product. [151]

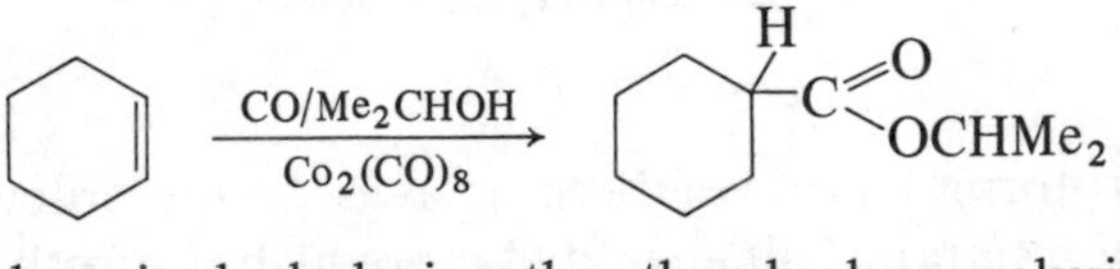

The final step here is alcoholysis rather than hydrogenolysis of the acyl-cobalt carbonyl intermediate.

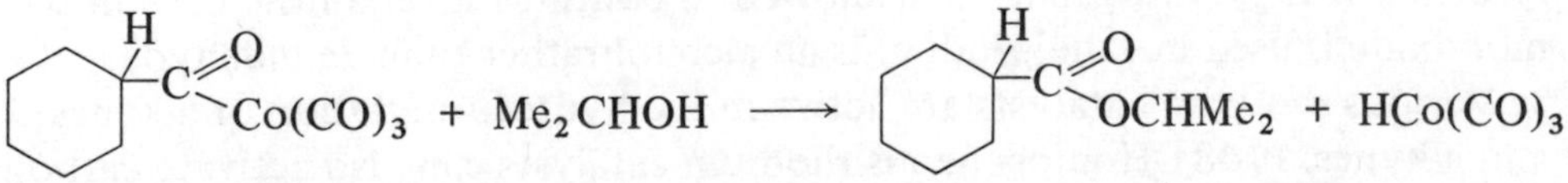

Similarly, incorporation of an amine rather than an alcohol yields the related amide. [60]

$$RCH=CH_2 + R'NH_2 + CO \xrightarrow{Co_2(CO)_8} RCH_2CH_2CONHR' + RCH \underset{CONHR'}{\overset{CH_3}{\diagdown}}$$

In a related reaction, the palladium chloride complex of an alkene can be treated with carbon monoxide in benzene to yield a β-chloroacid chloride. [152] In the presence of ethanol at about 100°C the product is an ethyl ester. [153]

$$>C=C< \xrightarrow[PdCl_2]{CO/benzene} >\overset{|}{C}-\overset{Cl}{\underset{|}{C}}-C\overset{O}{\diagdown_{Cl}}$$

$$>C=C< \xrightarrow[PdCl_2]{CO/EtOH} >\overset{|}{C}-\overset{H}{\underset{|}{C}}-C\overset{O}{\diagdown_{OEt}}$$

α,β-Unsaturated esters and β-diesters can also be prepared by slight variations of the reaction conditions. [154] Bis(triphenylphosphine–dichloropalladium is a particularly reactive catalyst for these carboxylations, allowing reaction at ambient temperatures. [155] Similar reactions are known for π-allyl palladium chloride complexes. [156]

$$HC\underset{CH_2}{\overset{CH_2}{\diagdown}}\!\!-Pd\!\!\underset{}{\overset{Cl}{\diagdown}} \quad\begin{matrix}\xrightarrow{CO/ROH} & CH_2=CH-CH_2CO_2R + Pd + HCl \\ \\ \xrightarrow{CO/benzene} & CH_2=CH-CH_2COCl + Pd\end{matrix}$$

The reaction of alkenes with carbon monoxide and water to give carboxylic acids can be carried out in a variety of ways, usually in the presence of an acid catalyst. Where no metal catalyst is employed, high pressures and temperatures are required.

$$>C=C< + CO + H_2O \xrightarrow{H_3O^\oplus} >\overset{|}{C}-\overset{H}{\underset{|}{C}}-C\overset{O}{\diagdown_{OH}}$$

With the addition of tetracarbonylnickel, milder conditions can be used, and the mechanism involves metal complex formation, a nickel-to-carbon rearrangement (CO insertion) followed by hydrolysis of the acyl nickel complex. [157]

87

$$Ni(CO)_4 \rightleftharpoons Ni(CO)_3 + CO$$

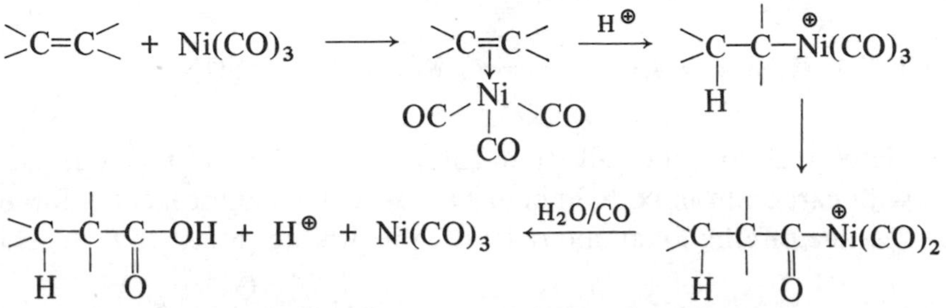

$$\ce{>C=C< + Ni(CO)_3 -> } \underset{\underset{CO}{\overset{Ni}{OC-\!|\!-CO}}}{\ce{>C=C<}} \xrightarrow{H^\oplus} \ce{>C-C-Ni(CO)_3}$$

$$\ce{>C-C-C-OH + H^\oplus + Ni(CO)_3} \xleftarrow{H_2O/CO} \ce{>C-C-C-Ni(CO)_2^\oplus}$$

Reaction conditions are even milder when an alkyne is used, the product then being an $\alpha,\beta$-unsaturated carboxylic acid (Reppe reaction). [158]

$$RC\equiv CH \xrightarrow[Ni(CO)_4]{CO/H_3O^\oplus} RCH=CH-CO_2H$$

## 5.4 Related syntheses of carbonyl compounds

### 5.4.1 *Cobalt and nickel carbonyl catalysts*

A new synthesis of symmetrical ketones involves the reaction of an alkyl or aryl mercury(II) halide with octacarbonyldicobalt in tetrahydrofuran. [159]

$$2RHgBr \xrightarrow[THF]{Co_2(CO)_8} \ce{R-C-R}\ (\ce{O})$$

Presumably an organocobalt compound is first formed with subsequent migration of the R group to give an acylcobalt intermediate as in the oxo reaction. A closely related reaction has been reported for tetracarbonylnickel. [160]

$$2PhHgX + Ni(CO)_4 \xrightarrow{DMF} PhCOPh + 2Hg + NiX_2 + 3CO$$

When an aryl iodide is heated with tetracarbonylnickel in tetrahydrofuran the product is the $\alpha$-diketone, resulting from a twofold migration reaction.

$$2PhI \xrightarrow[THF]{Ni(CO)_4} PhCOCOPh$$

By contrast, when aryl iodides or vinyl bromides are heated with tetra-carbonylnickel in an alcohol solvent, especially in the presence of the sodium alkoxide, the product is an ester. [161, 162]

$$PhI \xrightarrow[\text{MeOH}]{\text{Ni(CO)}_4} PhCO_2Me$$

### 5.4.2 Iron carbonyl catalysts

Reaction of pentacarbonyliron with sodium amalgam gives sodium tetra-carbonylferrate(-II), which has become a valuable reagent for the synthesis of aldehydes and ketones.

$$Fe(CO)_5 \xrightarrow[\text{THF}]{\text{Na—Hg}} Na_2Fe(CO)_4$$

The anion, $[Fe(CO)_4]^{2\ominus}$, is a powerful nucleophile and on reaction with a primary alkyl bromide yields an alkyl iron complex which on treatment with triphenylphosphine followed by acetic acid yields an aldehyde. The mechanism once again involves alkyl group migration from the metal atom to a carbonyl carbon atom. [163]

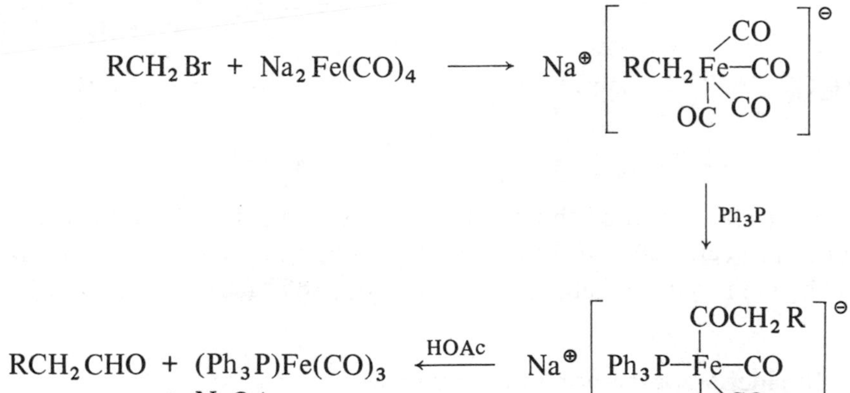

$$RCH_2Br + Na_2Fe(CO)_4 \longrightarrow Na^{\oplus} \left[ \begin{array}{c} CO \\ RCH_2Fe-CO \\ OC \ CO \end{array} \right]^{\ominus}$$

$$\Big\downarrow Ph_3P$$

$$RCH_2CHO + (Ph_3P)Fe(CO)_3 \xleftarrow{\text{HOAc}} Na^{\oplus} \left[ \begin{array}{c} COCH_2R \\ Ph_3P-Fe-CO \\ OC \ CO \end{array} \right]^{\ominus}$$
$$+ NaOAc$$

The anion $[RCH_2Fe(CO)_4]^{\ominus}$ can be converted by reaction with carbon monoxide into the acyl anion $[RCH_2COFe(CO)_4]^{\ominus}$. Either of these anions may be decomposed with halogens to yield an acyl halide and ultimately a carboxylic acid, ester or amide depending on the solvent. [164]

$$[RCH_2Fe(CO)_4]^{\ominus}$$
or
$$[RCH_2COFe(CO)_4]^{\ominus}$$

$$\xrightarrow{X_2} RCH_2COX$$

$$\xrightarrow{H_2O} RCH_2CO_2H$$
$$\xrightarrow{R'OH} RCH_2CO_2R'$$
$$\xrightarrow{R'R''NH} RCH_2CONR'R''$$

In the case of the anion $[RCH_2Fe(CO)_4]^{\ominus}$, treatment with halogen must initiate migration of the alkyl group to carbonyl carbon.

The anionic species, $[RCH_2Fe(CO)_4]^{\ominus}$ can also be allowed to react as a nucleophile with a second alkyl halide to give a neutral species which after rearrangement (carbonyl insertion) undergoes reductive elimination in the presence of a suitable co-ordinating solvent (see Section 2.5) to liberate a ketone. [165]

$$[RCH_2Fe(CO)_4]^{\ominus} \xrightarrow{R'CH_2X} \left[ \begin{array}{c} CH_2R' \\ | \\ RCH_2-Fe-CO \\ OC^{/} | \ ^{\backslash}CO \\ CO \end{array} \right]$$

$$\downarrow \text{solvent}$$

$$RCH_2COCH_2R' + (\text{solvent})Fe(CO)_3 \longleftarrow \left[ \begin{array}{c} O \ \ CH_2R' \\ \| \ \ | \\ RCH_2-C-Fe-CO \\ OC^{/} CO^{\backslash} \text{solvent} \end{array} \right]$$

The direct reaction of a lithium alkyl, RLi, with pentacarbonyliron gives the acyl iron species $Li^{\oplus}[RCOFe(CO)_4]^{\ominus}$, which can then be alkylated with an alkyl halide to yield a ketone as above [166, 167] (see also Section 7.7).

## 5.5  Organoborane–carbon monoxide reactions

It has been known since 1937 that diborane, $B_2H_6$, reacts with carbon monoxide at 100°C under 20 atmospheres pressure ($2 \times 10^3$ kN m$^{-2}$) to give an addition compound, $H_3BCO$. A useful extension of this reaction to organoboranes was reported by Brown in 1967, who found that a wide variety of trialkylboranes, $R_3B$, react readily with carbon monoxide in suitable solvents at 100–125°C and atmospheric pressure to give products of rearrangement formulated as polymers $(R_3CB=O)_x$. Oxidation of these intermediates affords the trialkylcarbinols in high yield. [30, 168]

$$R_3B + CO \longrightarrow (R_3CBO)_x \xrightarrow{H_2O_2} R_3COH$$

The trialkylboranes are in turn readily available from alkenes and diborane (see Sections 6.3 and 7.3). Further developments have extended this carbon monoxide insertion reaction to allow synthesis of primary and secondary alcohols as well as tertiary alcohols, and also aldehydes and ketones.

The reaction of trialkylboranes with carbon monoxide involves three successive migrations of an alkyl group from the boron atom to carbonyl carbon, and the migrations can be controlled by variations of reaction conditions and suitable choice of alkyl group attached to boron. Complete migration from boron to carbon leads eventually to the tertiary alcohol; migration of two alkyl groups leads variously to a secondary alcohol or a ketone, migration of only one alkyl group gives rise to a primary alcohol or aldehyde.

All these results are consistent with the following mechanisms of migration.

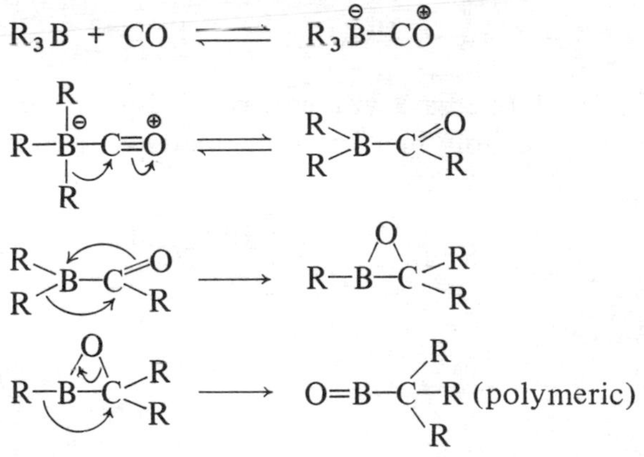

Each migration step amounts to the synthesis of a new carbon–carbon bond, one of the carbon atoms involved being that provided by the carbon monoxide. Subsequent transformation at each migration stage will now be discussed.

The polymers $(R_3CBO)_x$ can be oxidized directly with alkaline hydrogen peroxide; in some cases it is preferable to convert the polymer to a monomer by reaction with ethan-1,2-diol prior to oxidation.

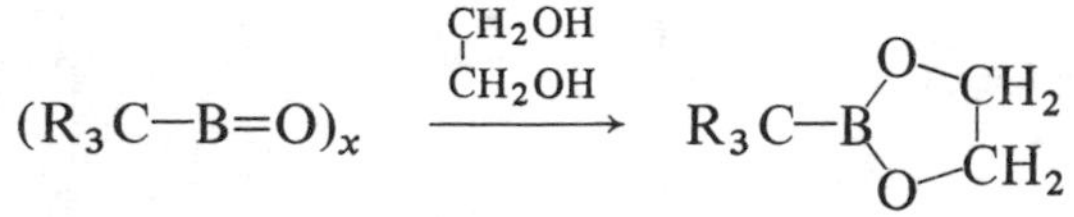

Oxidative cleavage of the carbon–boron bond leads to a tertiary alcohol; formation of carbon–oxygen from carbon–boron bonds will be discussed in Chapter 7, but can be shown here briefly as:

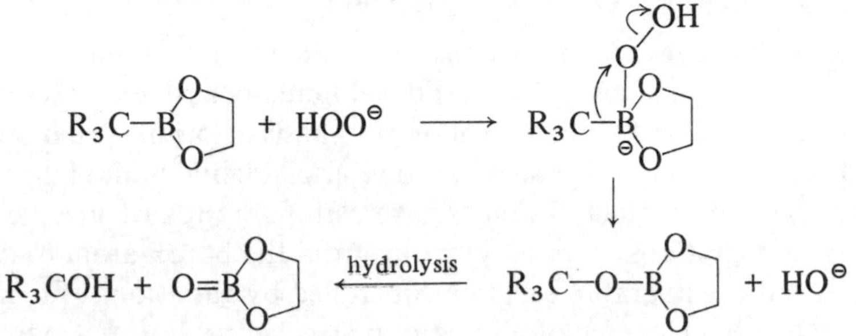

$$R_3C-B\langle\overset{O}{O}\rangle + HOO^{\ominus} \longrightarrow R_3C-B\langle\overset{O}{\underset{\ominus O}{}}\rangle \overset{OH}{}$$

$$R_3COH + O=B\langle\overset{O}{O}\rangle \xleftarrow{hydrolysis} R_3C-O-B\langle\overset{O}{O}\rangle + HO^{\ominus}$$

Migration of the third alkyl group from boron to carbon in the rearrangement process is slow and can be inhibited by the presence of water, possibly by conversion of the boraoxiran to a stable dihydroxy compound.

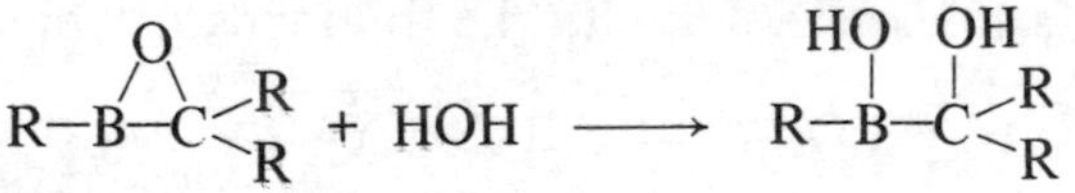

$$R-B\overset{O}{\underset{\triangle}{}}C\langle\overset{R}{R} + HOH \longrightarrow R-B\overset{HO}{\underset{}{}}\overset{OH}{\underset{}{}}C\langle\overset{R}{R}$$

Hydrolysis of this intermediate with aqueous sodium hydroxide then produces a secondary alcohol; the more usual oxidative hydrolysis produces a ketone.

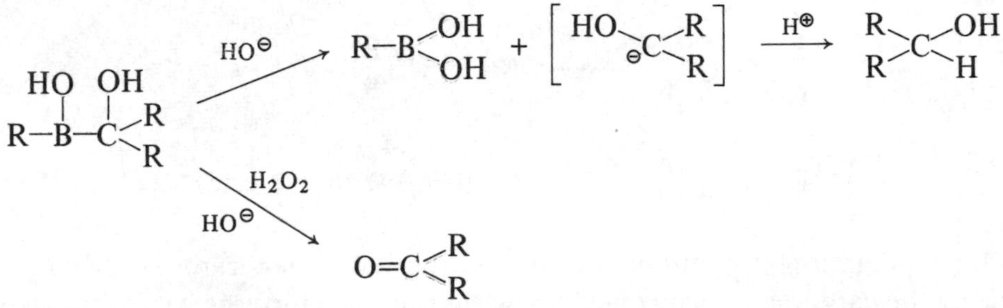

$$R-B\overset{HO}{\underset{}{}}\overset{OH}{\underset{}{}}C\langle\overset{R}{R} \xrightarrow{HO^{\ominus}} R-B\langle\overset{OH}{OH} + \left[HO\underset{\ominus}{}C\langle\overset{R}{R}\right] \xrightarrow{H^{\oplus}} R\overset{}{\underset{R}{}}C\langle\overset{OH}{H}$$

$$\xrightarrow[HO^{\ominus}]{H_2O_2} O=C\langle\overset{R}{R}$$

In the presence of lithium tetrahydridoborate a reductive process intervenes after one migration from boron to carbon has occurred, the product being an α-hydroxyalkyldialkylborane.

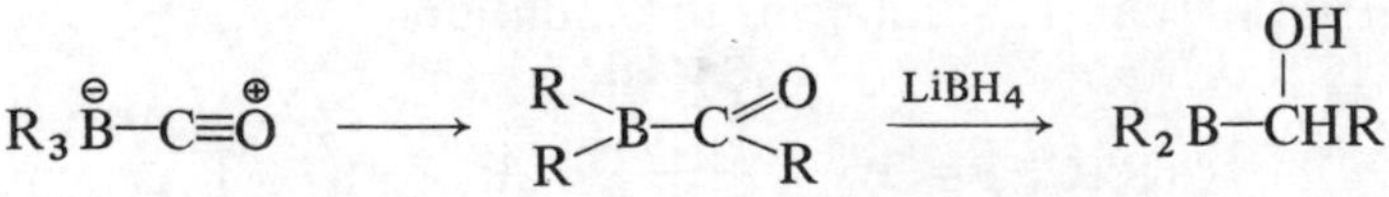

$$R_3\overset{\ominus}{B}-\overset{\oplus}{C}\equiv\overset{}{O} \longrightarrow \overset{R}{\underset{R}{}}B-C\langle\overset{O}{R} \xrightarrow{LiBH_4} R_2B-\overset{\overset{OH}{|}}{C}HR$$

Alkaline hydrolysis or oxidative hydrolysis as before then yields products as shown below.

$$R_2B-\overset{\overset{\displaystyle OH}{|}}{C}HR$$

$\xrightarrow{\text{HOH}}$ $R_2BOH + RCH_2OH$

$\xrightarrow{H_2O_2/HO^{\ominus}}$ $2ROH + RC-H$ 
$\overset{||}{O}$

In the above aldehyde synthesis only one of the three alkyl groups initially attached to boron is present in the desired product. A more economical conversion of an alkene into an aldehyde is provided by the use of the secondary borane, 9-borabicyclo[3.3.1]nonane, which on reaction with an alkene gives a tertiary borane. [169]

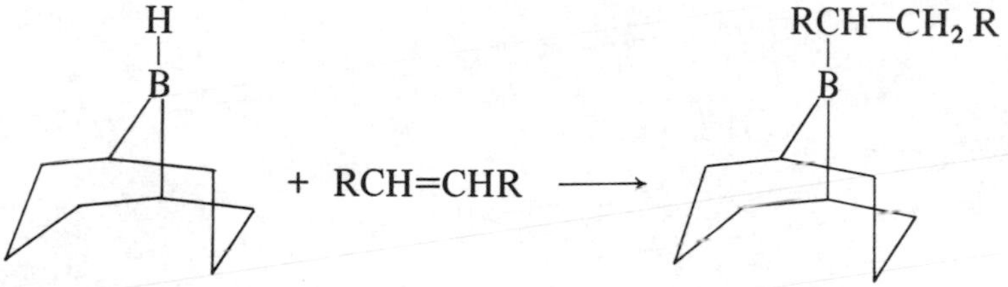

$+ RCH=CHR \longrightarrow$

On combination of this tertiary borane with carbon monoxide, migration of the newly introduced alkyl group occurs. This same group then appears in the aldehyde after oxidation. [170]

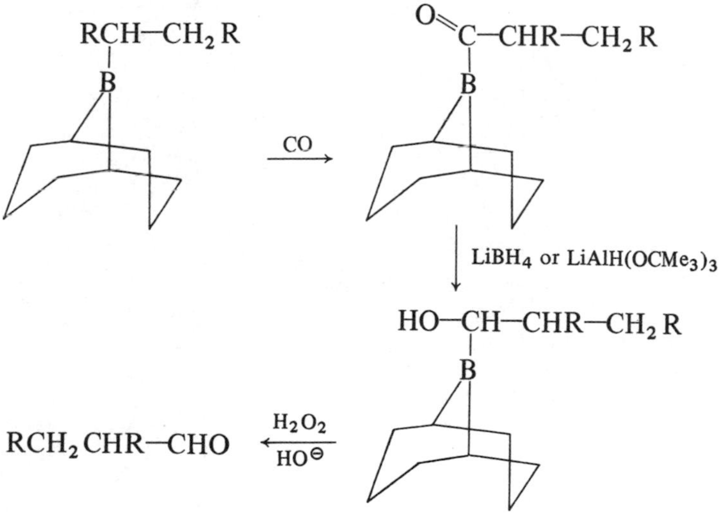

$\xrightarrow{CO}$

$\downarrow$ LiBH$_4$ or LiAlH(OCMe$_3$)$_3$

$RCH_2CHR-CHO$ $\xleftarrow[HO^{\ominus}]{H_2O_2}$

93

2,3-Dimethyl-2-butylborane ('thexyl borane', $RBH_2$) can be used for the synthesis of tertiary boranes $RR'R''B$ in which all three alkyl groups are different. Subsequent insertion of carbon monoxide followed by preferential migration of the groups $R'$ and $R''$ leads to the synthesis of unsymmetrical ketones. [171, 172]

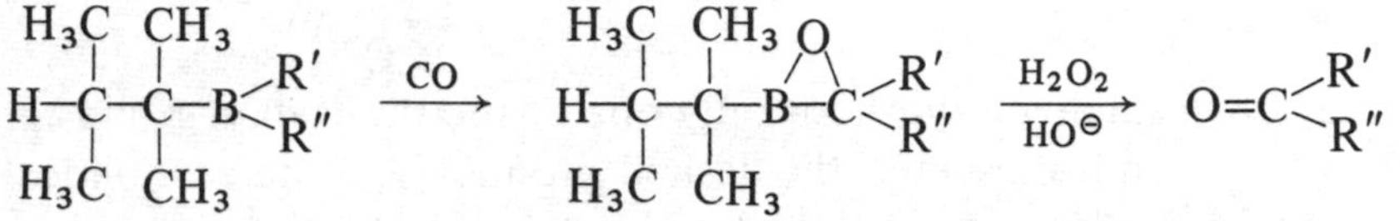

$$
\begin{array}{ccc}
\underset{\underset{H_3C}{|}}{\overset{\overset{H_3C}{|}}{H-C}}-\underset{\underset{CH_3}{|}}{\overset{\overset{CH_3}{|}}{C}}-B\diagdown^{R'}_{R''} & \xrightarrow{CO} & \underset{\underset{H_3C}{|}}{\overset{\overset{H_3C}{|}}{H-C}}-\underset{\underset{CH_3}{|}}{\overset{\overset{CH_3}{|}}{C}}-B-C\diagdown^{R'}_{R''} & \xrightarrow[HO^{\ominus}]{H_2O_2} & O=C\diagdown^{R'}_{R''}
\end{array}
$$

# Part III  Synthesis of bonds linking carbon to other atoms

# Synthesis of carbon–hydrogen bonds

**6**

## 6.1 Introduction

Metal hydrides, both neutral and anionic, have long been recognized as important reducing agents. The use of both stable and transient organometallic hydrides for the reduction of organic substrates is a logical extension of metal-hydride chemistry. Transient hydride intermediates are involved in catalytic hydrogenations using soluble organometallic catalysts, and these procedures are of much current interest. The very different hydrolytic cleavage of carbon–metal bonds is well known, and finds particular application in the preparation of hydrocarbons from halogeno compounds, in the formation of aldehydes from certain acyl metal carbonyl complexes, and in the process of hydrogenation via the addition of diborane or boron hydrides to alkenes and alkynes. Other aspects of hydroboration relevant to Part III are dealt with in Chapter 7 in relation to the synthesis of carbon–nitrogen and carbon–oxygen bonds. Reductive demercuration is becoming an important synthetic procedure; specific examples are given in Section 6.4 below and in Section 7.4.

## 6.2 Protolytic cleavage of carbon-metal bonds

As already mentioned in Section 2.7, organometallic derivatives of a variety of electropositive metals undergo solvolysis by water, alcohols or other proton sources to generate the corresponding hydrocarbon.

$$C_6H_5Li + H_2O \longrightarrow C_6H_6 + LiOH$$

$$RMgBr + MeOH \longrightarrow RH + BrMgOMe$$

$$(C_2H_5)_3Al + 3EtOH \longrightarrow 3C_2H_6 + Al(OEt)_3$$

$$RHgR + HCl \longrightarrow RH + RHgCl$$

$$ArCH_2SiCl_3 \xrightarrow{MeOH/KOH} ArCH_3$$

The use of a standard solution of methylmagnesium iodide to determine the number of moles of 'active' hydrogen in an unknown compound, as measured by the volume of methane liberated, is associated with the name of Zerevitinov. A synthetic application of the hydrolytic cleavage of Grignard reagents is found in the preparation of specifically deuterated compounds from the corresponding halogeno compounds.

$$RBr \xrightarrow{Mg} RMgBr \xrightarrow{D_2O} RD$$

Acid hydrolysis of organoboranes to yield hydrocarbons is an important aspect of hydroboration, discussed in Section 6.3 below.

The hydration of an alkyne to a carbonyl compound, catalysed by mercury(II) salts in acid solution, involves the synthesis both of a carbon–hydrogen and a carbon–oxygen bond.

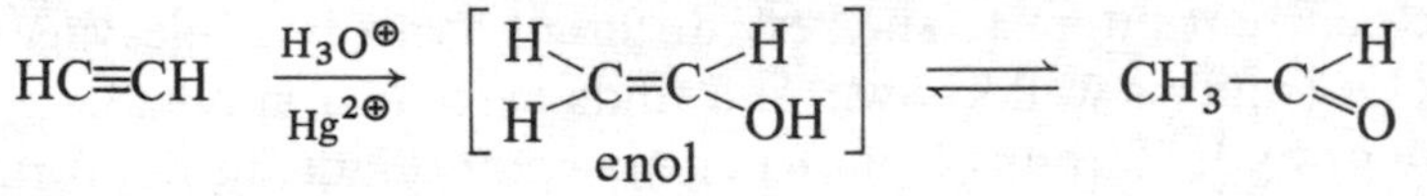

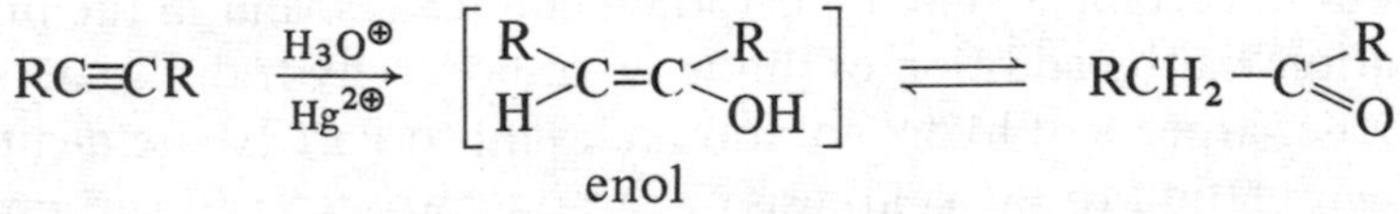

In these reactions the intermediate is an organomercury adduct which undergoes hydrolysis at the carbon–mercury bond to form the enol, tautomerically equivalent to the aldehyde or ketone.

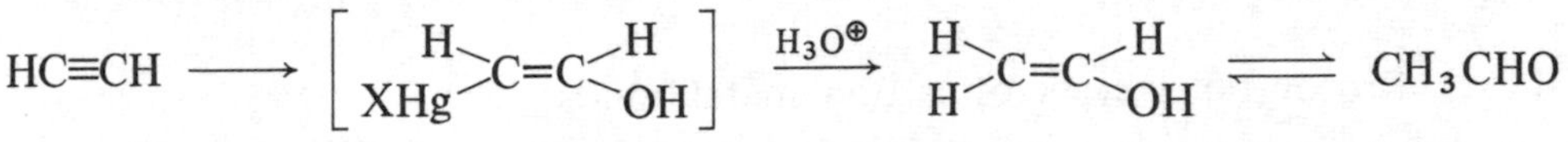

These and related processes are discussed in Sections 7.4.4 and 7.5.3 in the context of carbon–oxygen bond formation (c.f. also Sections 2.3 and 2.7).

## 6.3    Hydroboration of alkenes and alkynes

The addition of borane to an alkene, with subsequent manipulation of the product to give an isomeric alkene, an alkane, or an alcohol was noted in Section 4.2.2 and will be discussed further in Section 7.3. The formation of

98

an alkane from the intermediate organoborane involves hydrolytic cleavage of the carbon–boron bond of the kind already referred to in Section 6.2 above. It is worth noting that trialkylboranes are relatively stable to aqueous acids and bases but yield the hydrocarbon in acetic acid.

$$3RCH{=}CH_2 \xrightarrow{B_2H_6} (RCH_2CH_2)_3B$$

$$\xrightarrow{3CH_3CO_2H} 3RCH_2CH_3 + B(OCOCH_3)_3$$

The mechanism of this facile cleavage by acetic acid presumably involves a six-membered transition state.

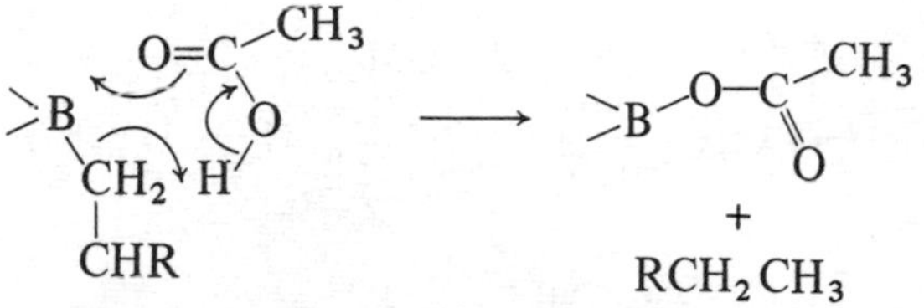

Hydroboration of an alkyne can lead to the corresponding alkene (see Section 4.1.1). [173] In the case of a terminal alkyne, the preferred method is to use a sterically-hindered alkyl borane such as bis(3-methyl-2-butyl)borane.

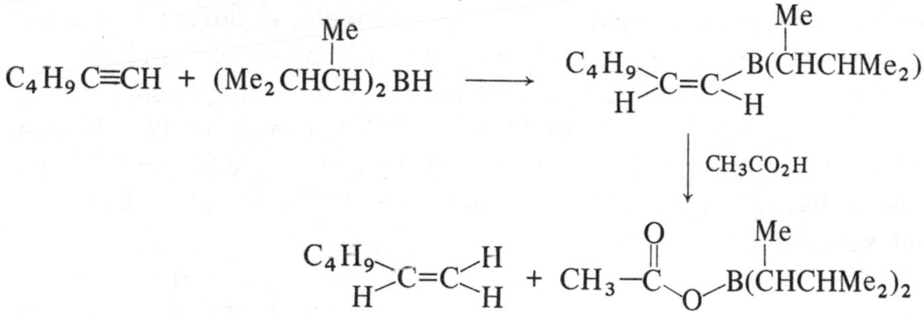

The related use of diisobutylhydridoaluminium as a reducing agent for alkynes has been discussed in Section 3.5.

## 6.4    Hydrogenolysis of carbon–metal bonds by hydrogen or metal hydrides

The trialkylboranes discussed in Section 6.3 above can be converted into an alkane and a dialkylborane by direct hydrogenation under pressure in the absence of any catalyst.

99

$$(C_4H_9)_3B \xrightarrow{H_2} C_4H_{10} + HB(C_4H_9)_2$$

The 'oxo' and related reactions involving the hydrogenolysis of an acyl–metal bond to give an aldehyde and a hydrido metal carbonyl have been discussed in Section 5.2. Hydrogenolyses of other carbon–metal bonds have also been reported and are likely to find increasing synthetic applications (cf. Sections 2.7 and 7.4). [174]

$$RPdCl \xrightarrow{LiAlH_4} RH$$

$$RHgBr \xrightarrow[H_2O]{Na-Hg} RH$$

$$\underset{\displaystyle \overset{OMe}{|}}{>C}\!-\!\underset{/\backslash}{C}\!-\!HgOAc \xrightarrow[(+H^\ominus)]{NaBH_4} \underset{\displaystyle \overset{OMe}{|}}{>C}\!-\!\underset{/\backslash}{C}\!-\!H + Hg + AcO^\ominus$$

$$RCu \xrightarrow{Bu_3PCuH} RH$$

## 6.5   Homogeneous hydrogenation using organometallic catalysts

Numerous organic complexes of metals such as rhodium, iridium, ruthenium, molybdenum and cobalt can act as homogeneous catalysts for catalytic hydrogenation. The most important of these is tris(triphenylphosphine)chloro-rhodium, a red crystalline compound known as Wilkinson's catalyst. In solution this complex undergoes oxidative addition of hydrogen (see Section 2.5) to form a metal hydride, from which one molecule of triphenylphosphine dissociates reversibly. [175–177]

$$(Ph_3P)_3RhCl \underset{}{\overset{H_2}{\rightleftharpoons}} (Ph_3P)_3\underset{\diagdown Cl}{\overset{\diagup H}{Rh{-}H}} \rightleftharpoons (Ph_3P)_2\underset{\diagdown Cl}{\overset{\diagup H}{Rh{-}H}} + Ph_3P$$

In the presence of a suitable alkene substrate an equilibrium is then established in which the alkene competes with triphenyl phosphine for co-ordination to the rhodium atom.

$$(Ph_3P)_2\underset{\diagdown Cl}{\overset{\diagup H}{Rh{-}H}} + RCH{=}CH_2 \rightleftharpoons (Ph_3P)_2\underset{\displaystyle \overset{|}{Cl}}{\overset{\displaystyle \overset{H}{|}}{Rh}}\underset{\displaystyle \underset{CH_2}{\parallel}}{\overset{\diagup H}{\diagdown CHR}}$$

Intramolecular hydride ion transfer from rhodium to carbon then gives rise to an alkyl rhodium complex still having one hydrogen atom attached to the rhodium atom. The initial hydride transfer is similar to that proposed for the hydroboration and hydroformylation of alkynes (see Sections 6.3, 7.3, and 5.2). The alkane is formed by a reductive elimination process (see Section 2.5) with regeneration of the catalyst.

$$
\underset{\substack{\text{Ph}_3\text{P} \\ \text{Ph}_3\text{P}}}{\overset{H}{\text{Rh}}} \cdots \overset{\text{H}}{\underset{\substack{\text{Cl} \\ \text{CH}_2}}{}} \text{CHR} \quad \xrightarrow[\text{cf. Section 2.6}]{\text{hydride migration}} \quad \underset{\substack{\text{Ph}_3\text{P} \\ \text{Ph}_3\text{P}}}{\overset{H}{\text{Rh}}} \underset{\text{Cl}}{\overset{}{\text{CH}_2}} \text{CH}_2\text{R}
$$

$$\downarrow \text{Ph}_3\text{P}$$

$$(\text{Ph}_3\text{P})_3\text{RhCl} + \text{CH}_3\text{CH}_2\text{R} \quad \xleftarrow[\substack{\text{of alkane} \\ \text{cf. Section 2.5}]{\text{reductive elimination}} \quad \underset{\substack{\text{Ph}_3\text{P} \\ \text{Ph}_3\text{P}}}{\overset{\text{Ph}_3\text{P} \quad H}{\text{Rh}}}\text{—CH}_2\text{CH}_2\text{R}$$

The proposed mechanism is consistent with the observation that undeuterated and dideuterated alkanes are formed from a mixture of $H_2$ and $D_2$ using homogeneous catalysts, whereas hydrogen–deuterium scrambling occurs with the use of heterogeneous catalysts.

$$\text{RCH=CH}_2 \quad \xrightarrow[\text{(Ph}_3\text{P)}_3\text{RhCl}]{H_2 + D_2 \text{ in benzene}} \quad \text{RCH}_2\text{CH}_3 \text{ and } \text{RCHDCH}_2\text{D}$$

Studies with deuterium have also shown that *cis*- addition to the double bond is preferred, and very little isomerization or hydrogen exchange occurs. In general, tris(triphenylphosphine)chlororhodium catalyses the hydrogenation of non-conjugated alkenes and alkynes at room temperature and atmospheric pressure. Conjugated alkenes and dienes which form intermediate chelates require higher pressures (60 atmospheres) for reduction. The rate of reduction of terminal alkenes is faster than that of internal alkenes and *cis*-alkenes are reduced more rapidly than the corresponding *trans*-alkenes. [178] Partial reduction of allenes is also possible and in the case of an unsymmetrical allene, the less substituted double bond is reduced. [179]

$$\text{RCH=C=CH}_2 \quad \xrightarrow[\text{(Ph}_3\text{P)}_3\text{RhCl}]{H_2/\text{benzene}} \quad \underset{H}{\overset{R}{>}}\text{C=C}\underset{H}{\overset{\text{CH}_3}{<}}$$

The main virtue of homogeneous organometallic catalysts lies in their ability to catalyse the hydrogenation of relatively unhindered carbon–carbon double bonds without affecting hindered double bonds or hydroxy, cyano, nitro, chloro, azo, oxido, or carbonyl functional groups. This selectivity is illustrated in the following examples. [180–182]

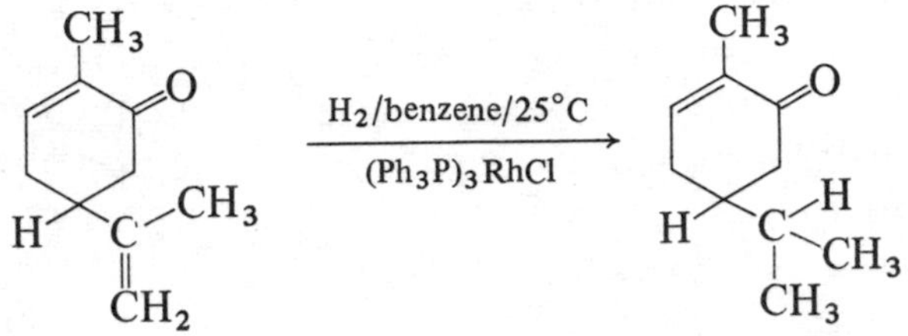

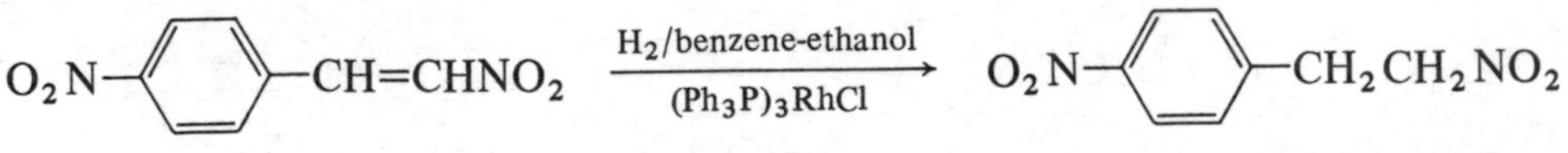

$$PhSCH_2CH=CH_2 \xrightarrow[\text{(Ph}_3\text{P)}_3\text{RhCl}]{\text{H}_2\text{/benzene}} PhSCH_2CH_2CH_3$$

In the last example, the unsaturated sulphur compound can be hydrogenated satisfactorily without the poisoning usually observed in heterogeneous catalysis.

Other useful catalyst systems include $\mu$-dichloro-$\pi$-benzene ruthenium, tris-(triphenylphosphine)trihydridocobalt, bis(triphenylphosphine)carbonylchloro-iridium, and bis(cyclopentadienyl)dihydridomolybdenum. [183]

The decarbonylation of aliphatic aldehydes by Wilkinson's catalyst also amounts to a synthesis of alkanes, with fission of a carbon–carbon bond and formation of a carbon–hydrogen bond (cf. Section 5.2).

6.6    Reduction of carbonyl compounds and of alkyl and aryl halides with organometallic hydrides

The reduction of carbonyl compounds to alcohols or alkanes and of halogeno-compounds to alkanes or arenes can be carried out very effectively by the use of hydride reducing agents, such as lithium tetrahydridoaluminate and sodium tetrahydridoborate. These reducing agents act as sources of nucleophilic hydride ions; organometallic intermediates are not formed and the reactions are therefore outside the scope of this book. However, various organometallic hydrides containing both carbon–metal and metal–hydrogen bonds have found increasing use in synthesis, especially in the reduction of carbonyl compounds and of alkyl and aryl halides.

Some examples are shown below. [184–188]

$$PhCOPh \xrightarrow[260°C]{Ph_2SiH_2} PhCH_2Ph$$

$$PhCH(OEt)_2 \xrightarrow[ZnCl_2]{Et_3SiH} PhCH_2OEt + Et_3SiOEt$$

$$>C=O \xrightarrow{Ph_2SnH_2} >C\overset{H}{\underset{OH}{\diagdown}} + (Ph_2Sn)_x$$

$$RX + (C_4H_9)_3SnH \longrightarrow RH + (C_4H_9)_3SnX$$
$$(X = Cl, Br, I)$$

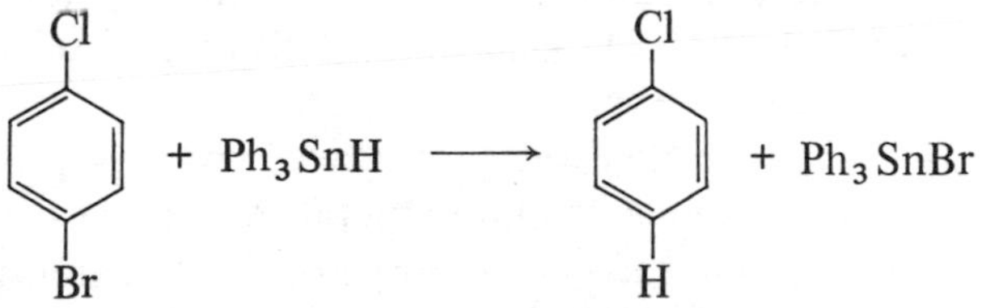

In general the above reactions are free radical in character [189] (cf. Section 2.4).

Similar reactions of organic derivatives of boron and aluminium hydrides have been reported.

# Synthesis of bonds linking carbon to nitrogen, phosphorus, oxygen, sulphur, and the halogens

# 7

## 7.1  Introduction

Organometallic compounds have long been of central importance in the construction of carbon–carbon bonds (Chapters 3, 4, and 5). Carbon–hydrogen bond formation was discussed in Chapter 6; the present chapter deals with the use of organometallic compounds or intermediates for the construction of bonds linking carbon to a variety of electronegative elements such as nitrogen, phosphorus, oxygen, sulphur, and the halogens. Many of these reactions have been reported only in the last few years, and their undoubted synthetic potential remains to be fully explored.

One process of wide generality and importance consists in the addition of a nucleophilic group X and an electrophilic metal-containing group $L_nM$ across a carbon–carbon double bond to give an organometallic intermediate which is then subjected to reduction with a reagent such as sodium tetrahydridoborate. In this second step the carbon–metal bond is replaced by carbon–hydrogen so that the overall result is the addition of HX across the double bond.

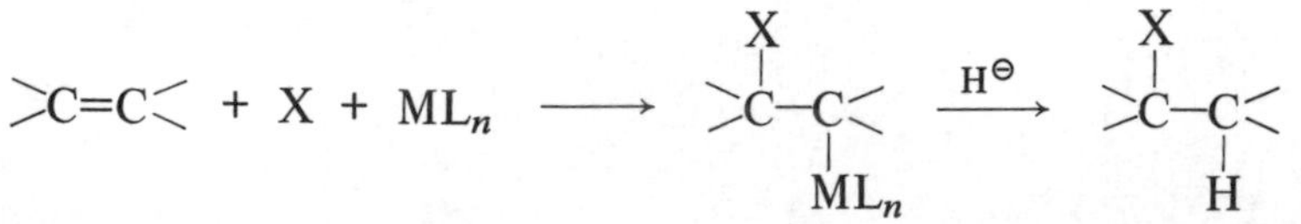

$$\text{>C=C<} + X + ML_n \longrightarrow \underset{\underset{ML_n}{|}}{\overset{\overset{X}{|}}{\text{>C–C<}}} \overset{H^{\ominus}}{\longrightarrow} \underset{\underset{H}{|}}{\overset{\overset{X}{|}}{\text{>C–C<}}}$$

Such reactions are treated here rather than in Chapter 6 since the feature of importance is formation of the C—X rather than the C—H bond; X in particular can be an oxygen or nitrogen atom and the metal commonly mercury, thallium or palladium.

## 7.2    Grignard reagents and related compounds

### 7.2.1    *Formation of carbon–nitrogen bonds*

Two reagents which react with organomagnesium compounds to give primary amines are *O*-methylhydroxylamine and chloramine.

$$RMgX + H_2NOCH_3 \longrightarrow RNH_2 + CH_3OMgX$$

$$RMgX + H_2NCl \longrightarrow RNH_2 + ClMgX$$

In both cases the nucleophilic carbanion, R:$^\ominus$, provided by the organo-metallic compound attacks the nitrogen atom with displacement of a methoxide or chloride ion. The method yields primary amines only, and is useful for the preparation of tertiary alkyl primary amines which cannot be obtained by reaction of ammonia with tertiary alkyl halides.

A new method for the synthesis of carbon–nitrogen bonds, leading again to the preparation of primary amines, consists in the reaction of two moles of a Grignard reagent with one mole of acetone oxime. [190]

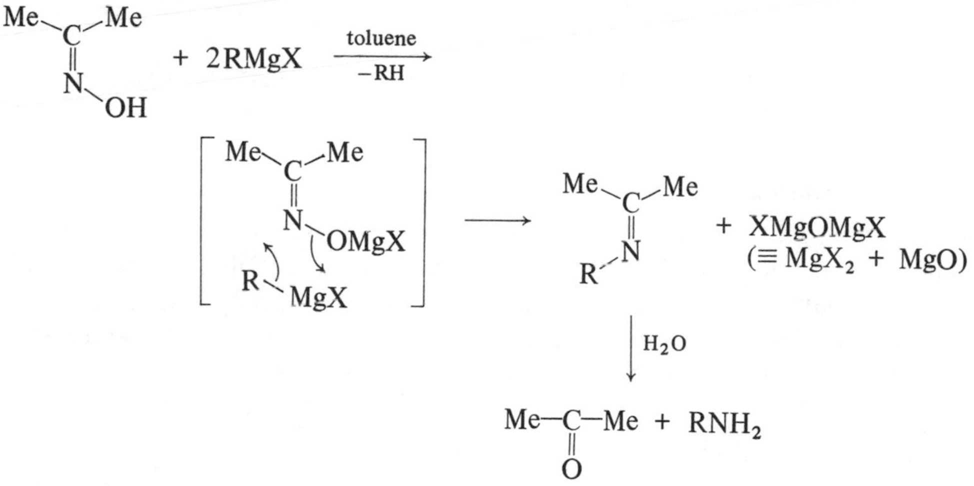

Grignard reagents react with aryl diazonium salts to yield azo compounds with formation of an aryl–nitrogen bond (cf. also Section 7.4.5).

$$ArN_2^{\oplus} X^{\ominus} + RMgX \longrightarrow ArN=N-R + MgX_2$$

### 7.2.2  *Formation of carbon–phosphorus bonds*

Many methods are known for the synthesis of organophosphorus compounds from organo–magnesium, lithium, and cadmium reagents. Various trivalent and pentavalent phosphorus halides or esters can be used, and the reactions can be classified as nucleophilic displacements, involving a carbanion provided by the organometallic reagent as nucleophile. [191]

$$PCl_3 \text{ or } P(OC_2H_5)_3 \xrightarrow{\text{3ArLi}} Ar_3P$$

$$2PCl_3 \xrightarrow{R_2Cd} 2RPCl_2$$

$$O{=}PCl_3 \xrightarrow{C_4H_9MgBr} (C_4H_9)_2P{\underset{OH}{\overset{O}{\diagdown}}}$$

$$R_2P{\underset{OC_4H_9}{\overset{O}{\diagdown}}} \xrightarrow{CH_2=CHMgBr} R_2P{\underset{CH=CH_2}{\overset{O}{\diagdown}}}$$

$$\underset{\underset{O}{\|}}{H{-}P(OC_2H_5)_2} \xrightarrow{RMgCl} \underset{\underset{O}{\|}}{H{-}PR_2}$$

Similar reactions are available for the synthesis of organic derivatives of arsenic, antimony, bismuth, boron, silicon, and many other elements.

### 7.2.3  *Formation of carbon–oxygen bonds*

Grignard reagents and related compounds react with oxygen to give the metal salt of a hydroperoxide, which can react with a second mole of organometallic reagent to give an alcoholate. Appropriate acidification can yield either a hydroperoxide or an alcohol, and the initial reaction with oxygen is almost certainly free radical in character [192] (cf. Section 2.4).

$$RMgX + O_2 \longrightarrow R{-}O{-}O{-}MgX \underset{\underset{RMgX}{\searrow}}{\overset{\overset{H_3O^{\oplus}}{\nearrow}}{}} \begin{matrix} ROOH \\ \\ 2ROMgX \xrightarrow{H_3O^{\oplus}} 2ROH \end{matrix}$$

$$CH_2=CH-CH_2-ZnCl \xrightarrow{O_2}$$

$$CH_2=CH-CH_2-O-O-ZnCl \xrightarrow{H_2O} CH_2=CH-CH_2-O-OH$$

A free radical chain process for the reaction of a Grignard reagent with oxygen is shown below, where In· is a free radical initiator.

Initiation:  $RMgX + In· \longrightarrow R· + In-MgX$

Propagation: $\begin{cases} R· + O_2 \longrightarrow R-O-O· \\ ROO· + RMgX \longrightarrow ROOMgX + R· \end{cases}$

The reaction of mixed trialkylalanes with oxygen to yield long chain primary alcohols is of considerable industrial importance, since these alcohols are intermediates in the synthesis of biodegradable detergents.

$$R_3Al + 1\tfrac{1}{2}O_2 \longrightarrow (RO)_3Al \xrightarrow[H_3O^\oplus]{H_2O \text{ or}} 3ROH$$

t-Butyl perbenzoate is a readily available peroxy compound and its reaction with either an alkyl or aryl Grignard reagent provides a useful synthesis of t-butyl ethers. [193]

$$C_6H_5-C\overset{O}{\underset{O-O-CMe_3}{\diagdown}} \xrightarrow{RMgX} C_6H_5-C\overset{O}{\underset{OMgX}{\diagdown}} + ROCMe_3$$

When R is an aryl group, subsequent hydrolysis of the t-butyl ether effects overall conversion of an aryl halide into a phenol.

The t-butyl perbenzoate reaction is sometimes shown as a carbanionic transfer involving a six-membered transition state:

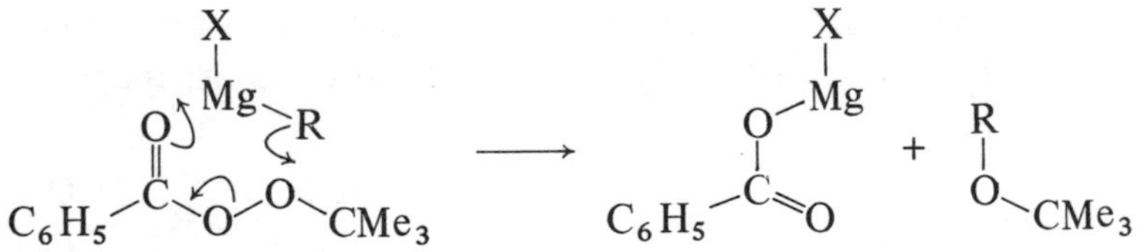

It seems more probable, however, that this reaction is also a free radical chain process:

Initiation: $\qquad$ In· + RMgX $\longrightarrow$ InMgX + R·

and/or $\quad$ In· + C$_6$H$_5$—C—OOCMe$_3$ $\longrightarrow$ In—OCMe$_3$ + C$_6$H$_5$—C—O·
$\qquad\qquad\qquad\qquad$ ‖ $\qquad\qquad\qquad\qquad\qquad\qquad\qquad\qquad$ ‖
$\qquad\qquad\qquad\qquad$ O $\qquad\qquad\qquad\qquad\qquad\qquad\qquad\qquad$ O

Propagation: $\begin{cases} \text{R· + C}_6\text{H}_5\text{—C—O—OCMe}_3 \longrightarrow \text{ROCMe}_3 + \text{C}_6\text{H}_5\text{—C—O·} \\ \qquad\qquad\quad \text{‖} \qquad\qquad\qquad\qquad\qquad\qquad\qquad \text{‖} \\ \qquad\qquad\quad \text{O} \qquad\qquad\qquad\qquad\qquad\qquad\qquad \text{O} \\ \text{C}_6\text{H}_5\text{—C—O· + RMgX} \longrightarrow \text{R· + C}_6\text{H}_5\text{—C—OMgX} \\ \qquad\quad \text{‖} \qquad\qquad\qquad\qquad\qquad\qquad\qquad \text{‖} \\ \qquad\quad \text{O} \qquad\qquad\qquad\qquad\qquad\qquad\qquad \text{O} \end{cases}$

### 7.2.4 *Formation of carbon–sulphur bonds*

Thiols and thioethers can be prepared from Grignard reagents and elemental sulphur.

$$\text{RMgX} \xrightarrow{S_8} \text{RS}_x\text{MgX} \xrightarrow{\text{RMgX}} \text{RSMgX} \begin{cases} \xrightarrow{H_3O^\oplus} \text{RSH} \\ \xrightarrow{\text{RMgX}} \text{RSR} + \text{Mg} + \text{MgX}_2 \\ \xrightarrow{R'I} \text{RSR}' + \text{MgXI} \end{cases}$$

Similar reactions are known for selenium and tellurium.

Some Grignard reagents are converted into thiol anions by reaction with thiirans (episulphides) such as the readily available propylene sulphide; the reaction is more general for alkyl and aryl lithium compounds. [194]

$$\text{RLi} + \text{MeCH—CH}_2 \longrightarrow \text{RS}^\ominus\text{Li}^\oplus + \text{MeCH=CH}_2$$
$$\diagdown\text{S}\diagup$$

The reaction of a Grignard reagent with sulphur dioxide is analogous to the well known reaction with carbon dioxide; the product is the magnesium salt of a sulphinic acid.

$$\text{RMgX} + \text{SO}_2 \longrightarrow \text{R—S}\underset{\text{OMgX}}{\overset{\text{O}}{\diagdown}} \xrightarrow{H_3O^\oplus} \text{R—S}\underset{\text{OH}}{\overset{\text{O}}{\diagdown}}$$

Similarly, the reaction of arylsulphonyl chlorides or esters with Grignard reagents gives rise to sulphones.

$$\text{RMgX} + \text{Ar}-\overset{\overset{\displaystyle O}{\|}}{\underset{\underset{\displaystyle O}{\|}}{S}}-\text{Cl} \longrightarrow \text{Ar}-\overset{\overset{\displaystyle O}{\|}}{\underset{\underset{\displaystyle O}{\|}}{S}}-\text{R} + \text{MgXCl}$$

### 7.2.5  *Formation of carbon–halogen bonds*

Reaction of Grignard reagents with iodine provides a useful synthesis of iodides from the corresponding chloro- or bromo-compounds,

$$\text{RMgX} + \text{I}_2 \longrightarrow \text{RI} + \text{IMgX}$$

The reaction is particularly useful when the normal halogen exchange reaction fails.

$$\text{Me}_3\text{C}-\text{CH}_2\text{Cl} \underset{\substack{\text{(i) Mg}\\\text{(ii) I}_2}}{\overset{\text{NaI}}{\diagdown\!\!\!\times}} \begin{array}{l}\text{Me}_3\text{C}-\text{CH}_2\text{I}\\[2em]\text{Me}_3\text{C}-\text{CH}_2\text{I}\end{array}$$

Many other organometallic compounds behave similarly. Thus treatment of a vinyl alane (see Section 3.5) with iodine gives the iodoalkene. [195]

$$\underset{H}{\overset{R}{>}}\text{C}=\text{C}\underset{\text{Al}(C_4H_9)_2}{\overset{H}{<}} \xrightarrow[\text{(ii) H}_3\text{O}^{\oplus}]{\text{(i) I}_2} \underset{H}{\overset{R}{>}}\text{C}=\text{C}\underset{I}{\overset{H}{<}}$$

## 7.3    Organoboranes

### 7.3.1  *Formation of carbon–oxygen bonds*

The hydroboration of alkenes with diborane (see Section 4.2.2) leads to trialkylboranes, which have considerable synthetic utility by virtue of the ability of various reagents to cleave the newly-formed carbon–boron bond(s). Of special importance are the reactions leading to alcohols, ketones, and aldehydes. [196, 197] (For the conversion of alkyl boranes into alkanes see Section 6.3.)

Conversion of a carbon–boron bond of an alkyl borane into a carbon–oxygen bond gives rise to a primary alcohol; in contrast, the direct acid-catalysed hydration of the alkene yields the isomeric secondary alcohol. Oxidation of the trialkylborane is achieved by reaction with alkaline hydrogen peroxide and the mechanism involves migration of a carbanionic R group from boron to oxygen.

$$R_3B \xrightarrow{HOO^{\ominus}} \quad \underset{R}{\overset{R}{\underset{|}{\overset{|}{B}}}}\!-\!O\!-\!O\!-\!H \longrightarrow \underset{R}{\overset{R}{>}}B\!-\!O\!-\!R + HO^{\ominus}$$

$$R_2BOH + ROH \xleftarrow{H_2O} \underset{R}{\overset{R}{>}}\overset{\ominus}{B}\underset{OR}{\overset{OH}{<}}$$

$$R_2BOH \xrightarrow[\text{HOO}^{\ominus} \text{ as above}]{\text{continued reaction with}} 2ROH + (HO)_3B$$

This oxidative rearrangement does not affect double or triple bonds, aldehydes, ketones, halides or nitriles. The R group migrates from boron to oxygen without skeletal rearrangement, and in the case of an optically-active R group the configuration of the carbon atom initially attached to the boron atom is retained in the product alcohol.

With an arylboronic acid, reaction with hydrogen peroxide gives rise to a phenol. [198]

$$ArB(OH)_2 \xrightarrow{H_2O_2} ArOH + B(OH)_3$$

Arylmercury compounds can also be converted into phenols via the arylborane. [199]

$$ArHgBr + R_2BH \longrightarrow R_2BAr \xrightarrow[HO^{\ominus}]{H_2O_2} ArOH + 2ROH$$

Hydroboration and subsequent oxidation of norbornene gives norbornyl alcohol consisting almost entirely of the *exo*-isomer. This result is a consequence of initial addition of the borane to the less-hindered face of the double bond.

110

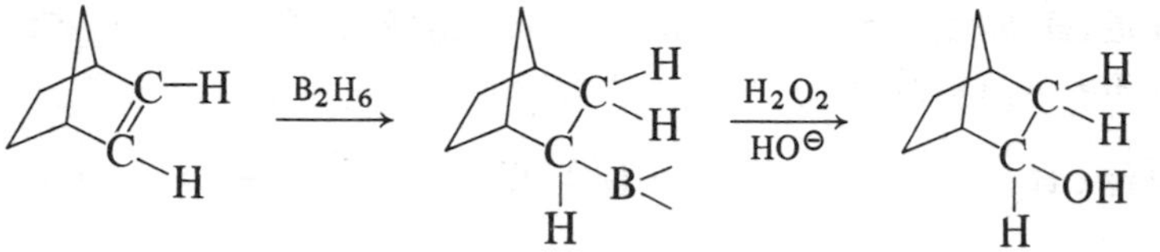

The oxidation of trialkylboranes with acidic dichromate gives carbonyl compounds, e.g.

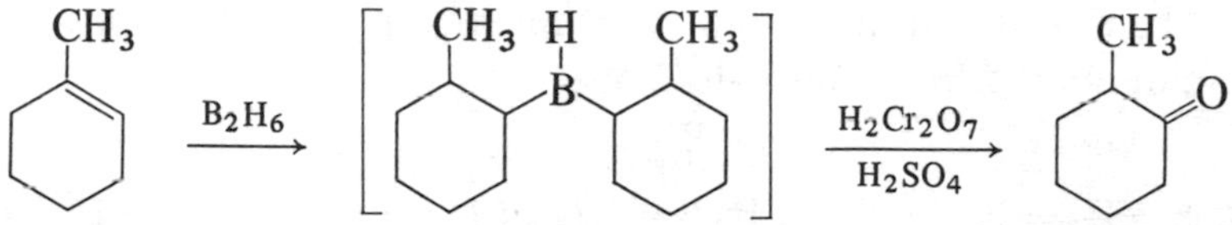

Hydroboration of a non-terminal alkyne followed by oxidative hydrolysis of the intermediate vinyl borane results in a useful preparation of ketones.

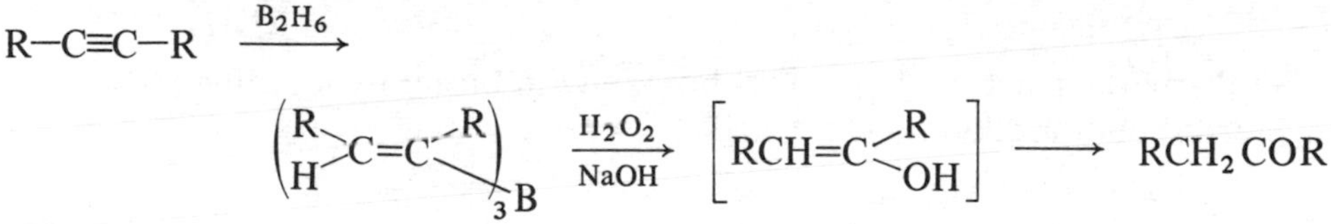

$$R-C\equiv C-R \xrightarrow{B_2H_6}$$

The formation of aldehydes from terminal alkynes requires the use of a sterically-hindered borane such as bis(3-methyl-2-butyl)borane (see Sections 4.1.1 and 6.3).

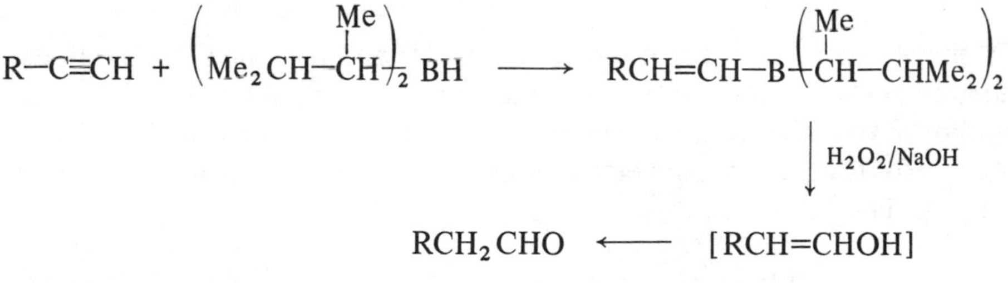

$$RCH_2CHO \longleftarrow [RCH=CHOH]$$

### 7.3.2 Formation of carbon–nitrogen, carbon–halogen and carbon–metal bonds

Alkyl boranes derived from the hydroboration of alkenes undergo ready reaction with either hydroxylamine-*O*-sulphonic acid ($H_3\overset{\oplus}{N}-OSO_3^{\ominus}$),

111

chloramine ($H_2NCl$), or dimethylchloramine ($Me_2NCl$) to afford the corresponding amine. [200]

$$RCH=CH_2 \xrightarrow{\ B_2H_6\ } (RCH_2CH_2)_3B \xrightarrow[H_3N^{\oplus}OSO_3^{\ominus}]{\ NH_2Cl\ or\ } RCH_2CH_2NH_2$$

$$\downarrow Me_2NCl$$

$$RCH_2CH_2NMe_2$$

The mechanism for amine formation is analogous to that proposed for the conversion of alkyl boranes into alcohols.

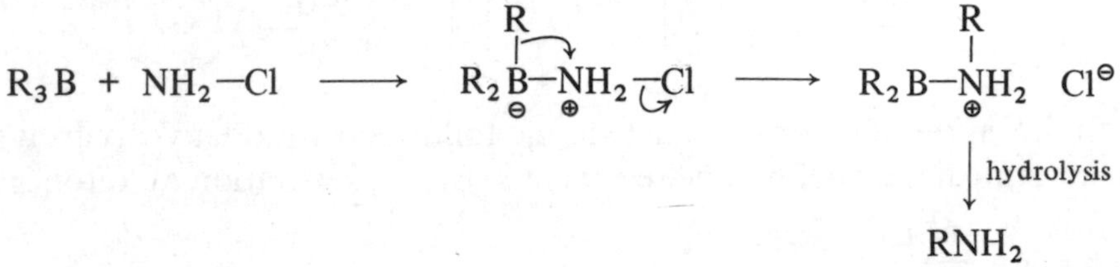

$$R_3B + NH_2-Cl \longrightarrow R_2\overset{\ominus}{B}-\overset{\oplus}{N}H_2-Cl \longrightarrow R_2B-\overset{\oplus}{N}H_2 \ Cl^{\ominus}$$

$$\downarrow \text{hydrolysis}$$

$$RNH_2$$

Similarly the reaction of an alkyl borane with bromine or iodine affords the corresponding alkyl halide. [201]

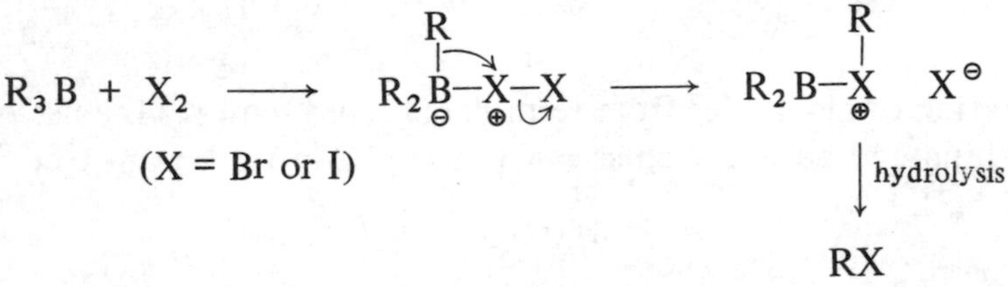

$$R_3B + X_2 \longrightarrow R_2\overset{\ominus}{B}-\overset{\oplus}{X}-X \longrightarrow R_2B-\overset{\oplus}{X} \ X^{\ominus}$$

$$(X = Br \text{ or } I) \qquad\qquad \downarrow \text{hydrolysis}$$

$$RX$$

In the above cases, as for the reaction of a trialkylborane with alkaline hydrogen peroxide, the configuration around the carbon atom which migrates from boron to a nitrogen, halogen or oxygen atom respectively is retained.

Arylboronic acids are converted by bromine or iodine into the corresponding aryl halide. [202]

$$ArB(OH)_2 \xrightarrow[H_2O]{\ X_2\ } ArX + B(OH)_3 + HX$$

Organoboranes show considerable promise for the synthesis of other organometallic compounds. Reaction of a trialkylborane with an aqueous suspension of lead(II) oxide yields the tetra-alkyl lead(IV) derivative and metallic lead.

112

$$2R_3B + 2PbO \xrightarrow[\text{NaOH}]{\text{H}_2\text{O}} R_4Pb + Pb + 2RB(OH)_2$$

In tetrahydrofuran solution, trialkylboranes react readily with mercury(II) acetate to give the corresponding acetoxymercury compound.

$$(RCH_2CH_2)_3B \xrightarrow[\text{THF}]{\text{Hg(OAc)}_2} 2RCH_2CH_2HgOAc$$

## 7.4 Organomercury compounds

### 7.4.1 *Aromatic mercuration*

Mercuration is a process in which a hydrogen atom attached to carbon is replaced by a mercury atom, and such a process is most favourable when the carbon atom is part of an aromatic ring. Thus, phenylacetoxymercury can be formed in good yield by the reaction of benzene with mercury(II) acetate in acetic acid at elevated temperatures, while thiophen reacts readily with mercury(II) chloride in a buffered solution at room temperature. [5]

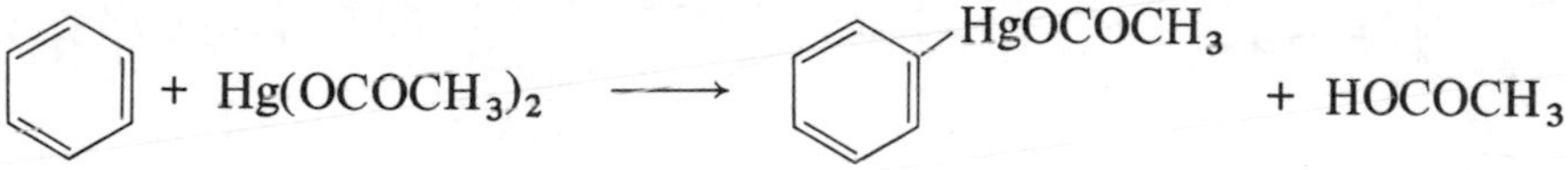

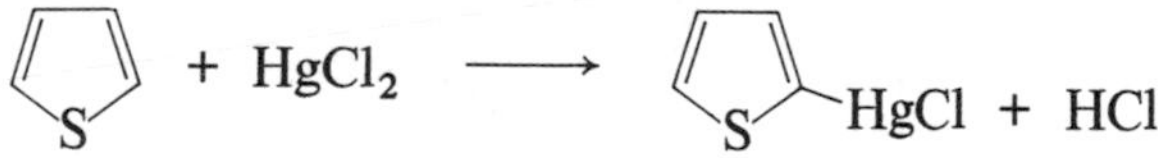

The mercuration reaction has wide generality and can also be carried out on hetero-aromatic and non-benzenoid aromatic compounds. These reactions are typical examples of electrophilic aromatic substitution and for acetoxymercuration the following mechanism is generally accepted.

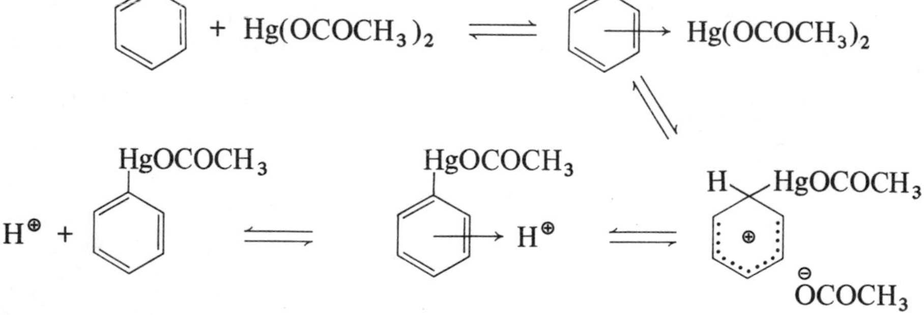

113

As with other electrophilic aromatic substitutions, groups such as $-CH_3$, $-OMe$, $-NR_2$, etc. activate the ring and direct the incoming acetoxymercury group to the *ortho-* and *para-*positions. Where co-ordination to the mercury is possible, a substituent may unexpectedly direct the electrophile to the *ortho-*position. Thus as already mentioned in Section 2.1, the mercuration of nitrobenzene occurs in the *ortho-*position. The same phenomenon occurs in thallation (see Section 7.5.1 below). In some cases mixtures of *meta-* and *para-*substituted products are obtained as a result of mobile mercuration-demercuration equilibria between the isomers.

## 7.4.2 *Formation of carbon–halogen bonds*

Arylmercury salts can be converted into aryl bromides and aryl iodides by reaction with bromine or iodine respectively (cf. Section 2.1). In the case of acetoxymercury compounds it is advantageous to exchange the acetate group with a halide ion prior to reaction with the halogen. [203]

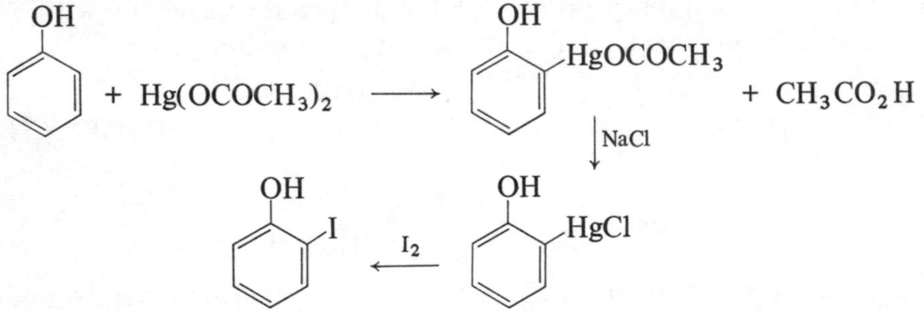

Polymercuration can lead to the formation of aryl polyhalides. [204]

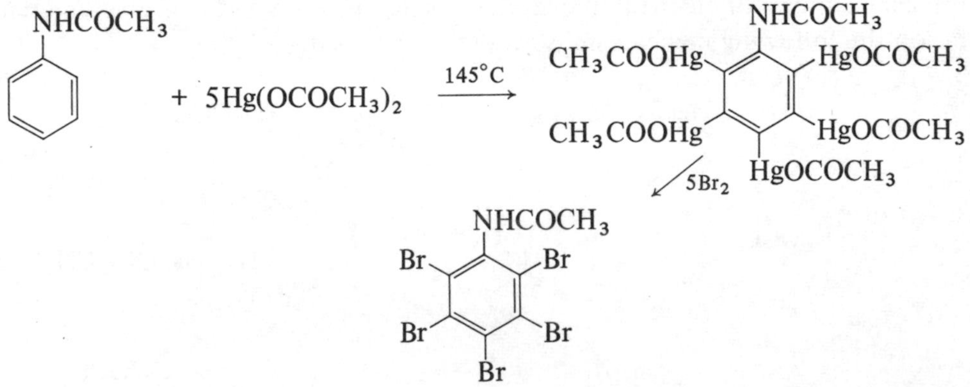

114

The mechanism for the substitution of mercury by halogen involves electrophilic attack, the leaving group being the mercury substituent rather than a proton (cf. Section 2.7).

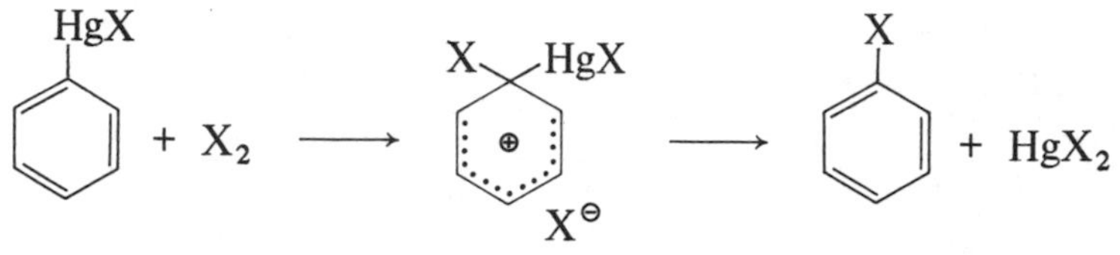

### 7.4.3 Oxymercuration of alkenes

An alcoholic solution of mercury(II) acetate reacts with alkenes in a conventional electrophilic addition process to form a product which contains a new carbon–oxygen bond and a new carbon–mercury bond. [4, 205]

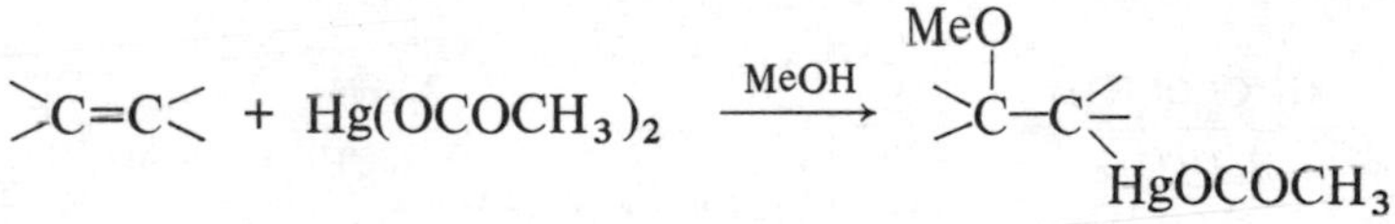

The stereochemistry of addition is commonly *trans* and the reaction is thought to proceed by electrophilic attack of the mercuriacetate cation to form an intermediate cyclic mercurinium ion, which undergoes subsequent attack by the solvent methanol (cf. Section 2.3).

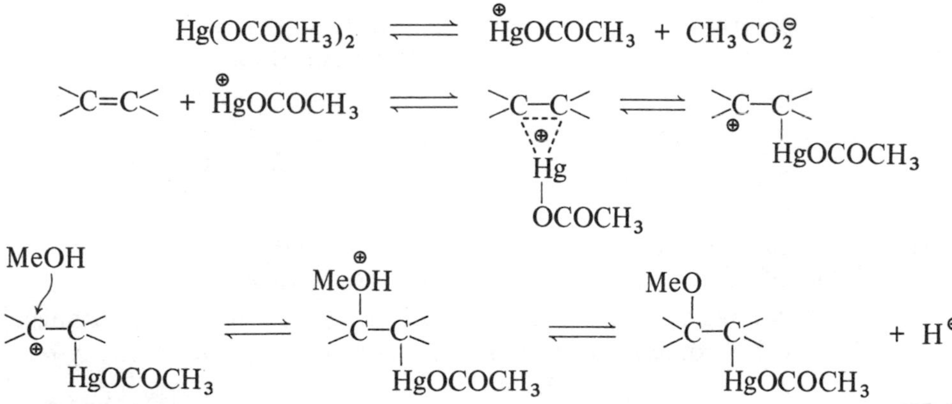

The oxymercuration process is reversible and deoxymercuration with regeneration of the alkene can be induced by halide ion.

*Cis*-oxymercuration has been found to occur in reaction with strained alkenes such as norbornene, and some variation in the above mechanism must apply here.

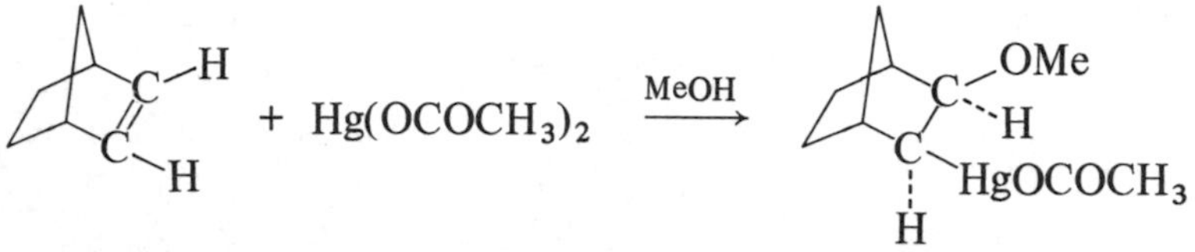

## 7.4.4  *Formation of carbon–oxygen and carbon–sulphur bonds*

This subsection deals with the synthesis of alcohols, ethers, thioethers, and esters. The products of alkene oxymercuration can be demercurated to give alcohols by treatment with alkaline sodium tetrahydridoborate. [206]

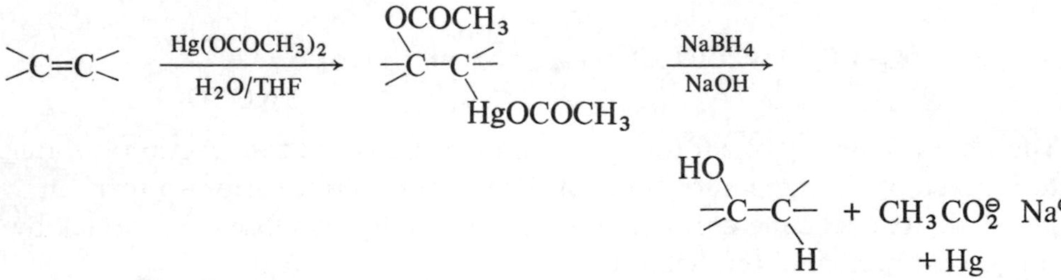

This demercuration process has formerly been regarded as a nucleophilic attack of a hydride ion on carbon, with displacement of metallic mercury and acetate ion.

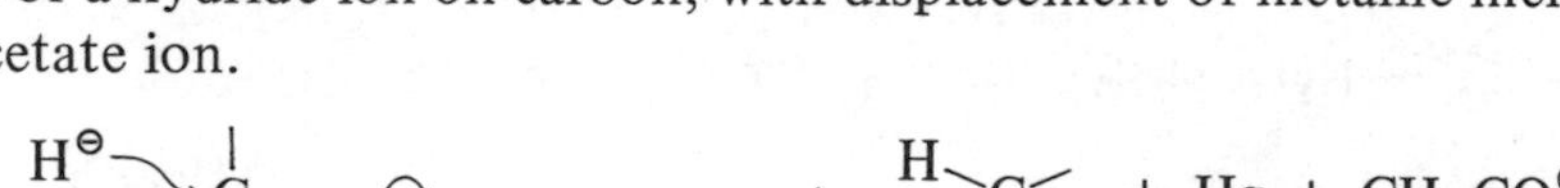

However it has now been established that the dermercuration of alkyl-mercury(II) halides by sodium tetrahydridoborate involves the intermediate formation of free radicals (cf. Sections 2.4 and 6.4). The detailed mechanism is not yet clear, but some kind of hydridomercury species could initially be formed. [207]

$$R-HgX \xrightarrow{NaBH_4} R-Hg-H \longrightarrow R\cdot + \cdot HgH$$

$$R\cdot + \cdot HgH \longrightarrow RH + Hg$$

The overall result is hydration of the alkene in the Markovnikov sense.

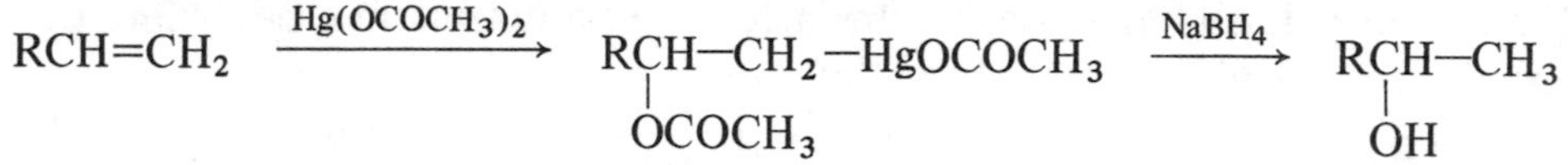

$$RCH{=}CH_2 \xrightarrow{Hg(OCOCH_3)_2} \underset{\underset{OCOCH_3}{|}}{RCH{-}CH_2{-}HgOCOCH_3} \xrightarrow{NaBH_4} \underset{\underset{OH}{|}}{RCH{-}CH_3}$$

Yields are practically quantitative for terminal and disubstituted internal alkenes, but variable for trisubstituted alkenes. The reaction displays high sensitivity to steric factors.

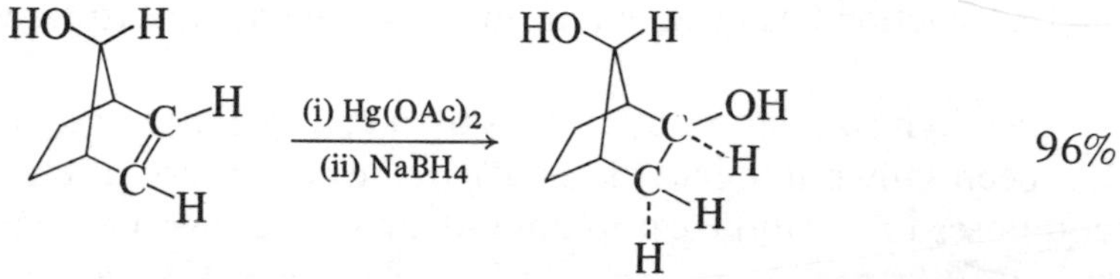

$$\text{(i) Hg(OAc)}_2 \qquad \text{(ii) NaBH}_4 \qquad 96\%$$

If the above reaction is carried out in alcohol rather than water, ethers are generally formed provided that the alcohol is reasonably nucleophilic. Ether formation is favoured by the use of mercury(II) trifluoroacetate and this reagent allows the preparation even of $t$-butyl ethers. [208]

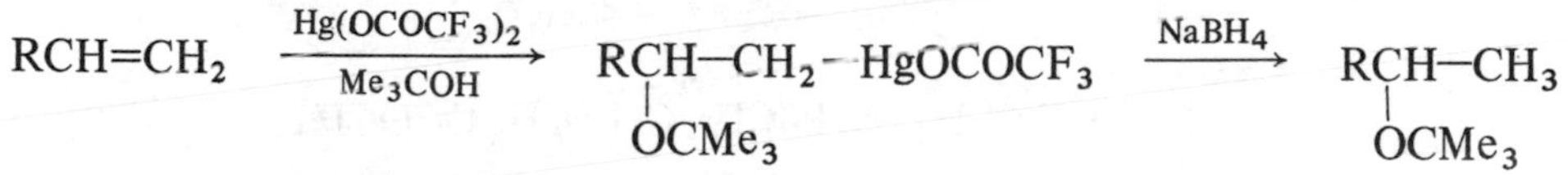

$$RCH{=}CH_2 \xrightarrow[Me_3COH]{Hg(OCOCF_3)_2} \underset{\underset{OCMe_3}{|}}{RCH{-}CH_2{-}HgOCOCF_3} \xrightarrow{NaBH_4} \underset{\underset{OCMe_3}{|}}{RCH{-}CH_3}$$

Oxymercuration of appropriate unsaturated alcohols can lead to five- or six-membered cyclic ethers by intramolecular nucleophilic attack of the hydroxyl group on the reacting double bond. [209]

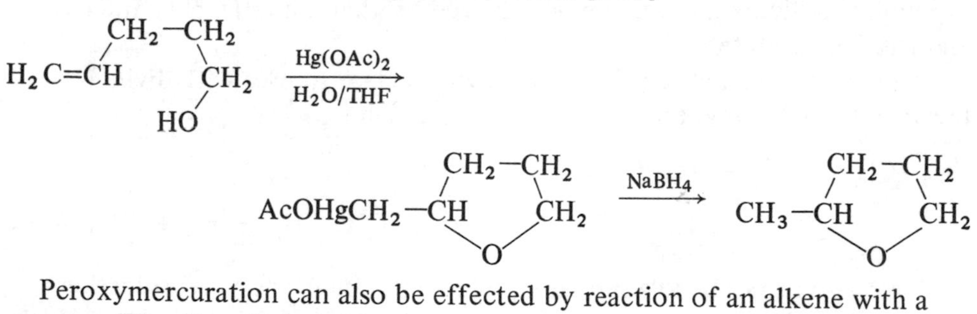

Peroxymercuration can also be effected by reaction of an alkene with a mercury(II) salt and $t$-butyl hydroperoxide. [210]

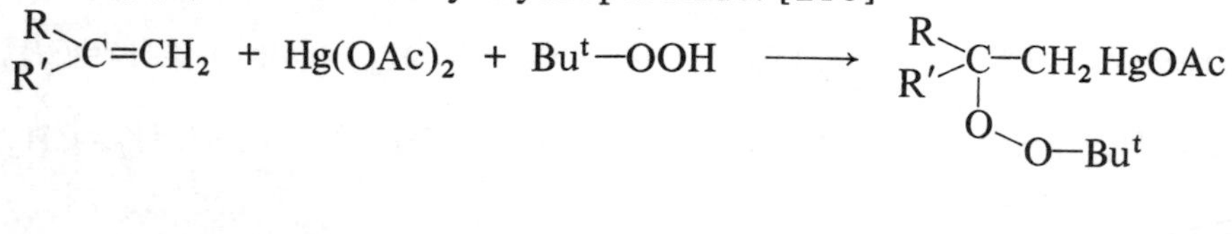

117

Demercuration with sodium tetrahydridoborate affords a mixture of a dialkylperoxide and an epoxide, the latter predominating in cases where the initial terminal alkene is disubstituted.

$$\begin{array}{c} R \\ R' \end{array}\!\!>\!\!C\text{--}CH_2 HgOAc \xrightarrow{\text{NaBH}_4} \begin{array}{c} R \\ R' \end{array}\!\!>\!\!C\text{--}CH_3 \quad + \quad \begin{array}{c} R \\ R' \end{array}\!\!>\!\!C\text{--}CH_2$$

(first and second structures bearing $O\text{--}O\text{--}Bu^t$ group; third an epoxide O)

The reaction has been extended to include $\alpha,\beta$-unsaturated ketones and esters.

The allylic oxidation of alkenes by mercury(II) acetate to form allyl acetates has been shown to occur via an allylacetoxymercury intermediate which decomposes to an allylic carbonium ion, metallic mercury, and acetate ion. [211]

$$PhCH_2 CH=CH_2 + Hg(OCOCH_3)_2 \longrightarrow PhCH=CH\text{--}CH_2 HgOCOCH_3$$

$$PhCH=CH\text{--}CH_2 HgOCOCH_3 \rightleftharpoons [PhCH\text{--}\overset{\oplus}{CH}\text{--}CH_2] + Hg + CH_3 CO_2^{\ominus}$$

$$\diagup CH_3CO_2{}^{\ominus}$$

$$\underset{\underset{OCOCH_3}{|}}{PhCH\text{--}CH=CH_2} + PhCH=CH\text{--}CH_2 OCOCH_3$$

$$40\% \qquad\qquad 60\%$$

Similar allylic oxidations can be achieved by thallium(III), palladium(II) and lead(IV) acetates.

In a possibly related reaction, carboxylic acids can react with divinylmercury to yield vinyl esters, mercury, and ethene.

$$CH_2=CH\text{--}Hg\text{--}CH=CH_2 + RCO_2 H \longrightarrow$$

$$RCO_2 CH=CH_2 + Hg + CH_2=CH_2$$

Phenols and thiophenols similarly yield vinyl ethers and vinyl thioethers respectively. [212]

$$CH_2=CH\text{--}Hg\text{--}CH=CH_2 + ArOH \longrightarrow ArOCH=CH_2 + Hg + CH_2=CH_2$$

$$CH_2=CH\text{--}Hg\text{--}CH=CH_2 + ArSH \longrightarrow ArSCH=CH_2 + Hg + CH_2=CH_2$$

Isonitriles can be readily converted into ureas and urethanes by their reaction with mercury(II) acetate in the presence of water, alcohols or primary amines. [213]

$$R-NC + H_2O \xrightarrow{Hg(OAc)_2} RNHCONHR$$

$$R-NC + R'OH \xrightarrow{Hg(OAc)_2} RNHCOOR'$$

$$R-NC + R'NH_2 \xrightarrow{Hg(OAc)_2} RNHCONHR'$$

The mercury(II) salt effects oxymercuration of the isonitrile, producing an iminoacetoxymercury compound with a new carbon–oxygen bond.

$$R-NC + Hg(OAc)_2 \longrightarrow R-N=C{<}^{HgOAc}_{OAc}$$

Subsequent migration of a co-ordinated solvent molecule from the mercury atom to the imine carbon atom is postulated. Concomitant cleavage of the carbon–mercury bond produces metallic mercury and an organic intermediate, which is transformed rapidly into the urea or urethane, depending on the solvent.

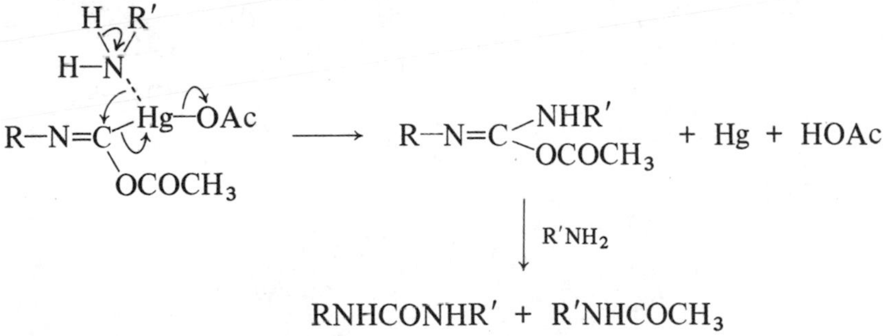

$$R-N=C{<}^{NHR'}_{OCOCH_3} + Hg + HOAc$$

$$\downarrow R'NH_2$$

$$RNHCONHR' + R'NHCOCH_3$$

### 7.4.5  Formation of carbon–nitrogen bonds

Water and alcohols are not the only solvents in which mercury(II) acetate can react with alkenes. The term 'solvomercuration' has been suggested as a suitably general term to encompass many reactions of a variety of mercury(II) salts in solvents such as acetonitrile or amines. The reaction of mercury(II) nitrate with terminal or cyclic alkenes in acetonitrile, followed by demercuration with alkaline sodium tetrahydridoborate, yields N-alkylacetamides which may be further hydrolysed to primary amines. [214]

119

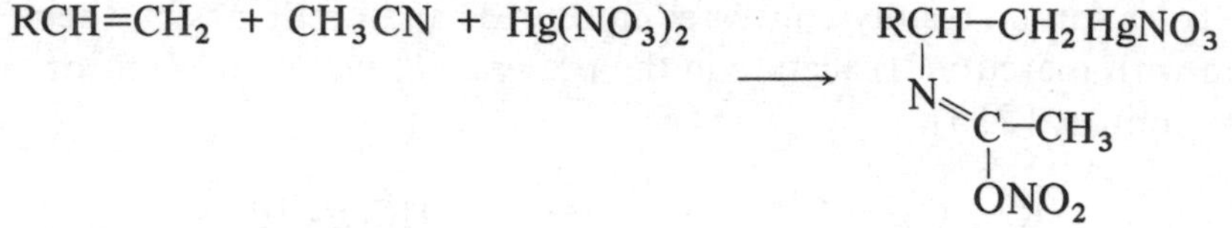

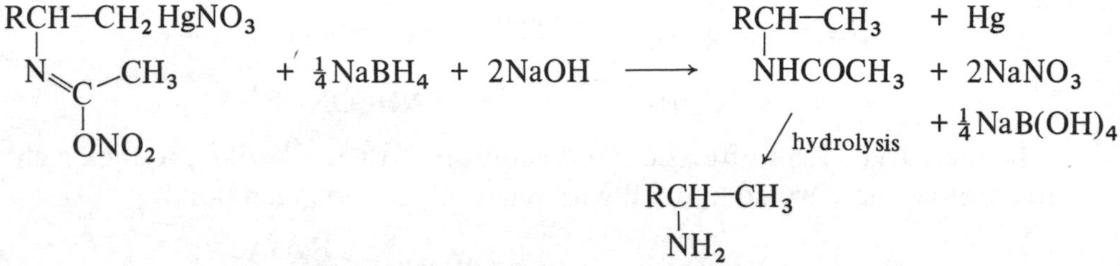

$$RCH{=}CH_2 + CH_3CN + Hg(NO_3)_2 \longrightarrow \underset{\underset{ONO_2}{\underset{|}{C{-}CH_3}}}{\overset{\overset{RCH{-}CH_2 HgNO_3}{|}}{N}}$$

$$\underset{\underset{ONO_2}{\underset{|}{C}}}{\overset{\overset{RCH{-}CH_2 HgNO_3}{|}}{N}}{-}CH_3 + \tfrac{1}{4}NaBH_4 + 2NaOH \longrightarrow \underset{NHCOCH_3}{\overset{RCH{-}CH_3}{|}} + Hg + 2NaNO_3$$

$$+ \tfrac{1}{4}NaB(OH)_4$$

hydrolysis

$$\underset{NH_2}{\overset{RCH{-}CH_3}{|}}$$

Primary or secondary amines can act as nucleophiles in the solvomercuration reaction of alkenes. Subsequent demercuration of the organometallic intermediates affords tertiary amines. [215, 216]

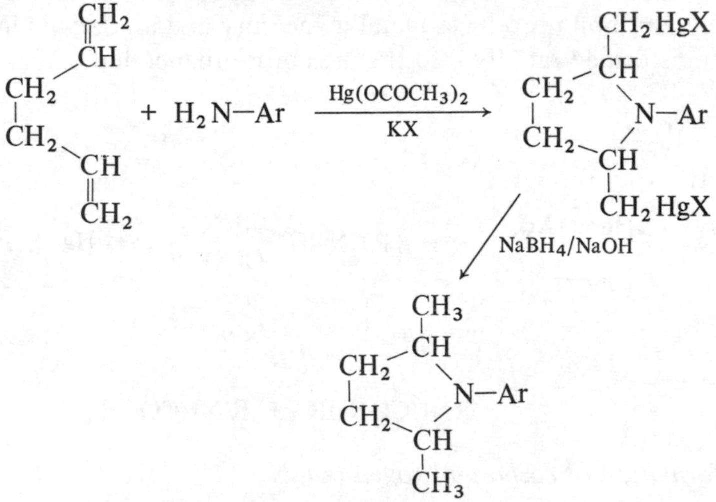

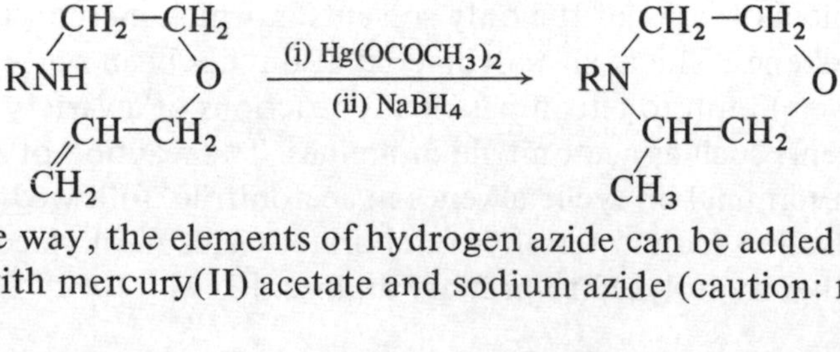

In the same way, the elements of hydrogen azide can be added to an alkene by reaction with mercury(II) acetate and sodium azide (caution: mercury(II)

120

azide is extremely hazardous), followed by demercuration of the resulting organomercury compound with sodium tetrahydridoborate. The reaction is successful only with terminal alkenes or strained cyclic alkenes. [217, 218]

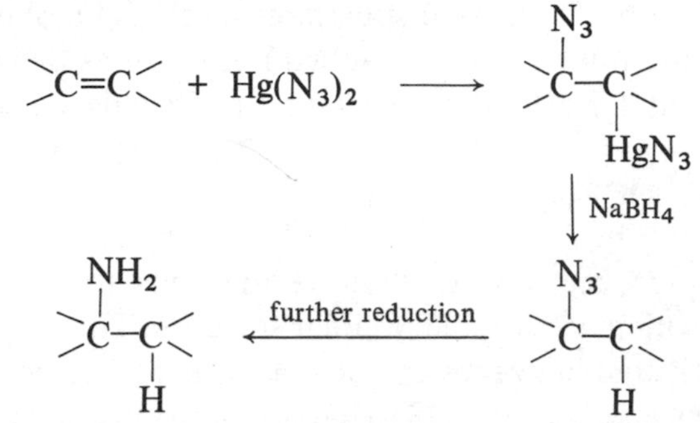

$$>C=C< \ + \ Hg(N_3)_2 \ \longrightarrow \ >\underset{HgN_3}{\overset{N_3}{C}}-C<$$

$$\downarrow NaBH_4$$

$$>\underset{H}{\overset{NH_2}{C}}-C< \quad \xleftarrow{\text{further reduction}} \quad >\underset{H}{\overset{N_3}{C}}-C<$$

Aryl diazonium salts react with organomercury compounds to yield unsymmetrical azo compounds (cf. Section 7.2.1). [219]

$$ArN_2^{\oplus} \, X^{\ominus} \ + \ R_2Hg \ \longrightarrow \ ArN{=}NR \ + \ RHgX$$

### 7.4.6 Oxidation of organomercury compounds

The oxidation of alkylmercury compounds with ozone also breaks the carbon—mercury bond. Primary alkyl groups are converted into carboxylic acids, secondary groups into ketones and tertiary into alcohols. [220]

$$(RCH_2)_2Hg \quad \xrightarrow[-76°C]{O_3} \quad RCO_2H$$

$$R_2CH{-}HgCl \quad \xrightarrow[10°C]{O_3} \quad RCOR$$

$$R_3C{-}HgCl \quad \xrightarrow[10°C]{O_3} \quad R_3COH$$

## 7.5 Organothallium compounds

### 7.5.1 Aromatic thallation

Thallium(III) acetate displays Lewis acid properties which enable it to function as a catalyst in electrophilic aromatic substitution reactions, particularly bromination. [221] Thallation (analogous to mercuration) of the aromatic ring does not occur with this reagent, but is successful with the more

121

electrophilic thallium(III) trifluoroacetate. [222] Substituted benzenes can then be prepared by cleavage of the carbon–thallium bond with formation of bonds linking carbon to nitrogen, oxygen, sulphur or halogen atoms.

Aromatic thallation of activated substrates (such as halobenzenes) is rapid at room temperature but deactivated substrates (such as benzoic acid) require the action of heat under reflux for several hours. Yields are high in all cases.

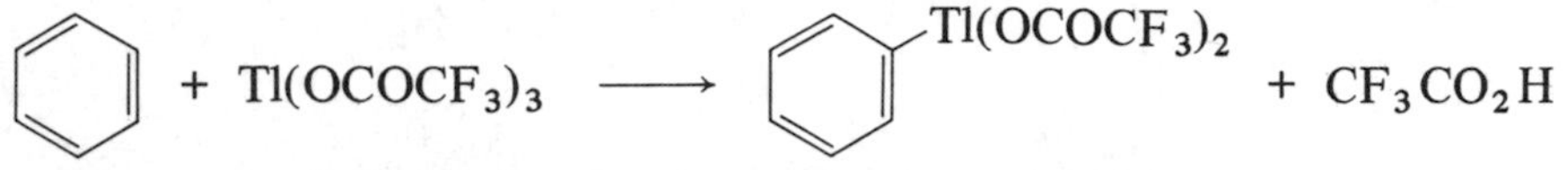

+ Tl(OCOCF3)3
Tl(OCOCF3)2
+ CF3CO2H

The aryl di(trifluoroacetoxy) thallium compounds are usually stable, colourless solids, which are soluble in polar solvents. The orientation of thallation in substituted benzenes can be controlled to some extent by the reaction conditions employed. [223] Under conditions of thermodynamic control, *meta* substitution is favoured, whereas kinetic control leads preferentially to *para* substitution, or *ortho* substitution in cases where the substituent can co-ordinate with the thallium reagent.

CH2CH2CH3
73°C
Tl(OCOCF3)3
room temperature
CH2CH2CH3
Tl(OCOCF3)2
CH2CH2CH3
Tl(OCOCF3)2

CO2Me
Tl(OCOCF3)3
CO2Me
Tl(OCOCF3)2

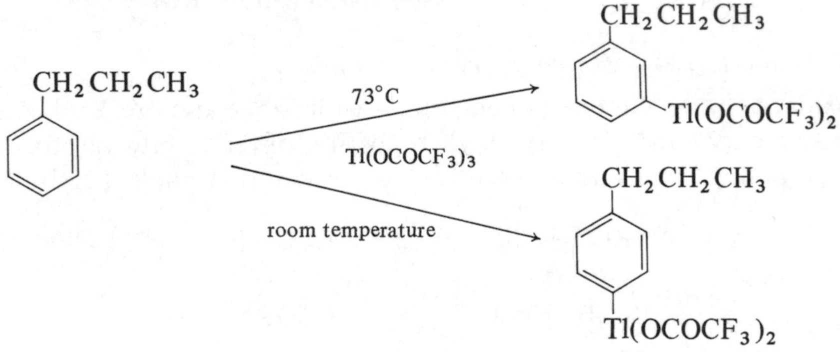

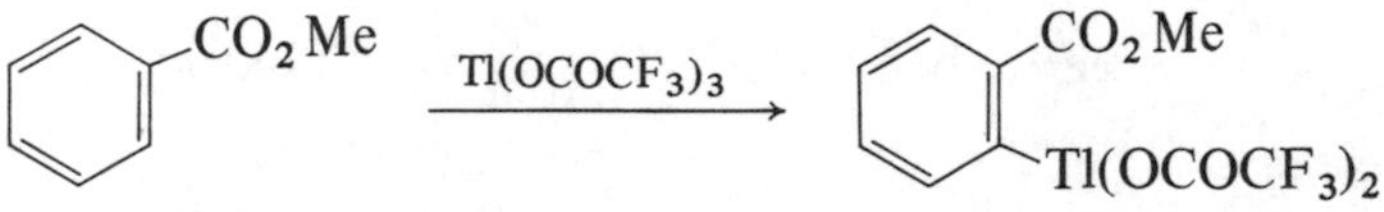

## 7.5.2 *Formation of carbon–nitrogen bonds*

Aniline can be produced by the photolysis of phenyl di(trifluoroacetoxy) thallium in the presence of ammonia and the process may be general. [24]

Tl(OCOCF3)3
Tl(OCOCF3)2
hν
NH3
NH2

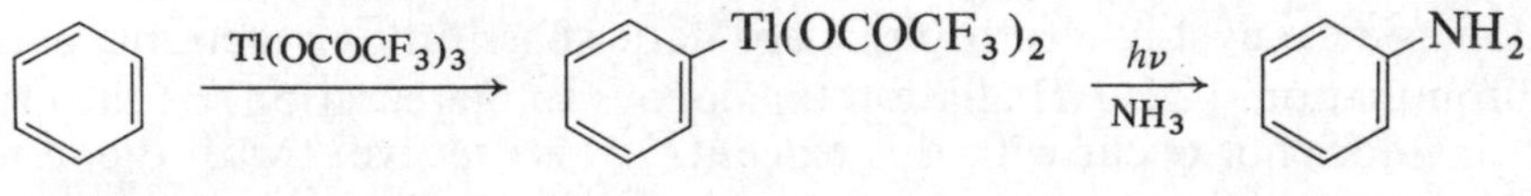

Aromatic nitroso compounds can be prepared by the reaction of aryl di-(trifluoroacetoxy) thallium compounds with nitrosyl chloride. [224]

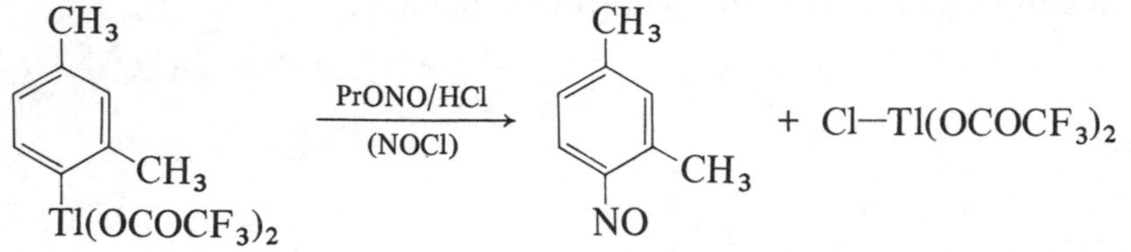

### 7.5.3 Formation of carbon–oxygen bonds

Phenols can be synthesized by oxidation of aryl di(trifluoroacetoxy) thallium compounds with lead tetra-acetate, followed by addition of triphenylphosphine and hydrolysis of the resulting aryl trifluoroacetate. The isolation of intermediates in the reaction sequence is not required. [225]

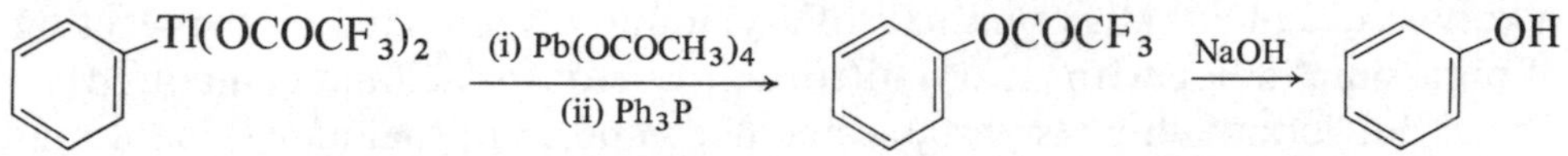

The oxidation of *para*-substituted aryl di(trifluoroacetoxy) thallium compounds with peroxytrifluoroacetic acid affords *para*-quinones, the reaction taking place with either elimination or migration of the *para*-substituent. [226]

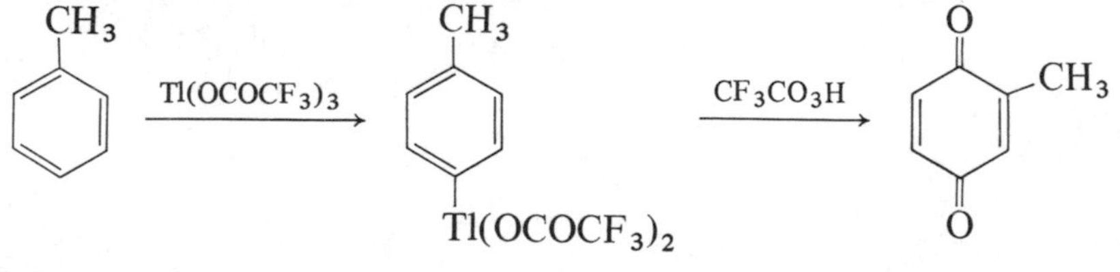

Carbon–oxygen bonds are also formed in oxythallation, dealt with in Section 7.5.6 below.

### 7.5.4 Formation of carbon–sulphur bonds

The irradiation of phenyl di(trifluoroacetoxy) thallium in an aqueous solution of potassium thiocyanate leads to the formation of phenyl thiocyanate. [227]

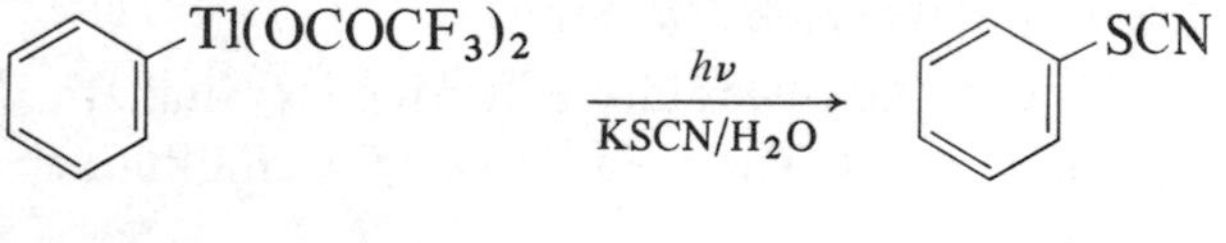

123

A more complicated reaction sequence is required for the synthesis of thiophenols. The critical step involves photolysis of an aryl bis(dithiocarbamato) thallium compound to give an aryl dithiocarbamate.

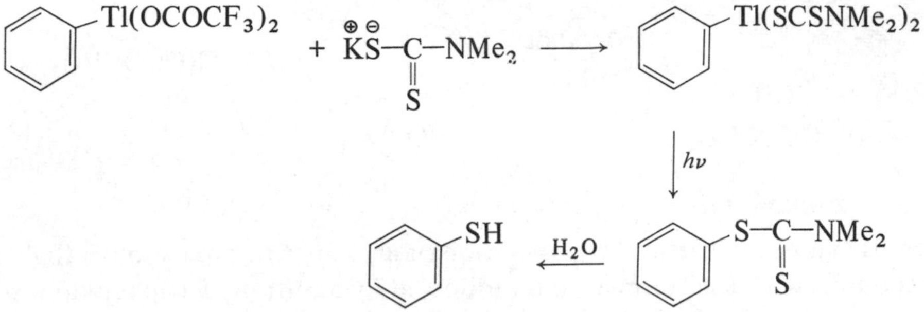

### 7.5.5 Formation of carbon–halogen bonds

A very easy and effective synthesis of aryl iodides is provided by the reaction of potassium iodide with an aryl di(trifluoroacetoxy) thallium compound. The aryldi–iodothallium is unstable and decomposes to thallium(I) iodide and the aryl iodide. [228]

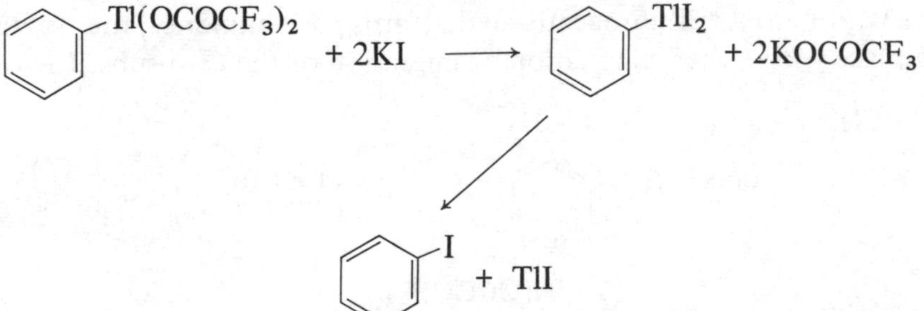

Various phenylthallium(III) salts have been converted into chlorobenzene in moderate yield by reaction with either copper(I) or copper(II) chloride. [229]

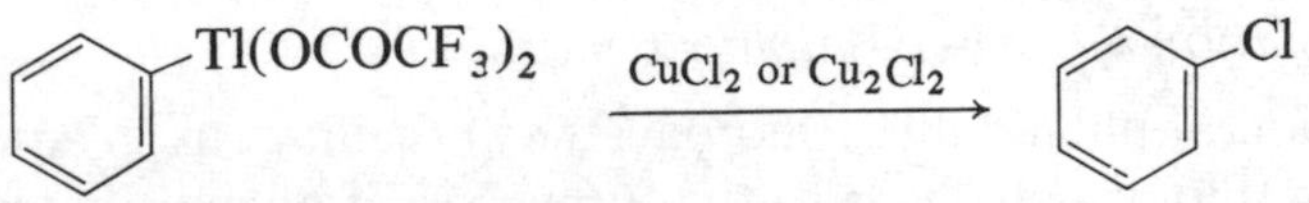

### 7.5.6 Oxythallation of alkenes and alkynes

Thallium(III) nitrate is capable of rapidly effecting oxythallation of carbon–carbon double and triple bonds. The resulting organothallium compound is

124

highly unstable and can undergo decomposition in a variety of ways, depending on the substituents present. This cleavage is thought to be heterolytic, resulting in the development of carbonium-ion character at the carbon atom of the original carbon–thallium bond (cf. Section 2.3).

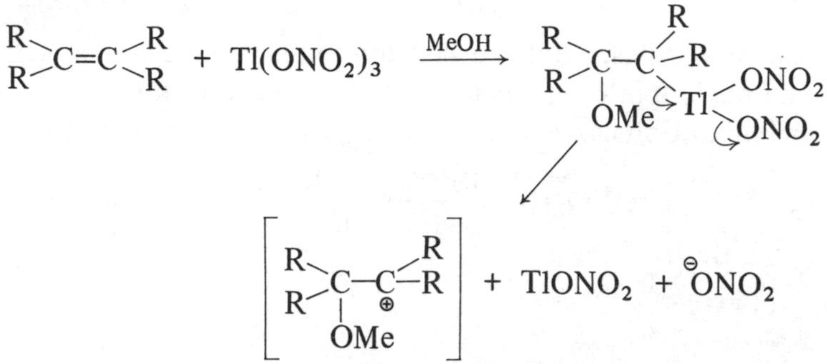

The carbonium ion commonly rearranges before capturing a solvent nucleophile. For example, cyclic alkenes undergo ring contraction, the final product being an aldehyde. [230]

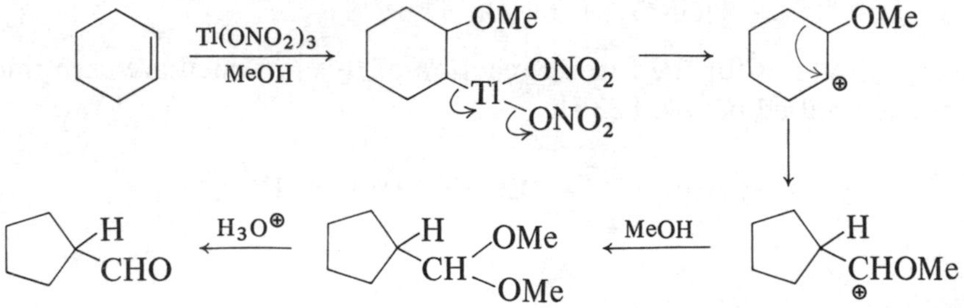

Oxythallation of styrenes leads to benzyl ketones, the intermediate carbonium ion being stabilized by phenyl migration.

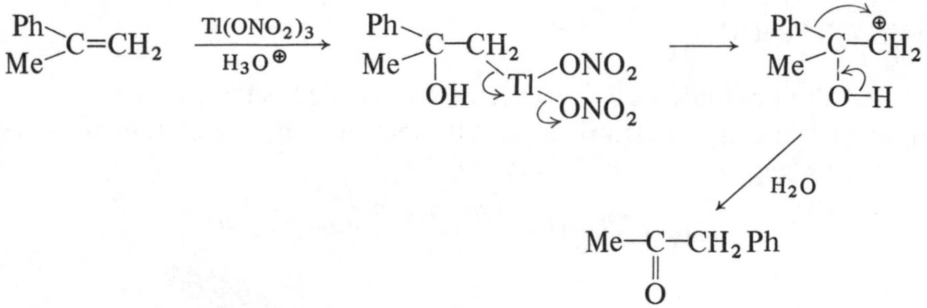

125

In similar fashion, methyl phenyl ketone is converted into an ester of phenylacetic acid. [231]

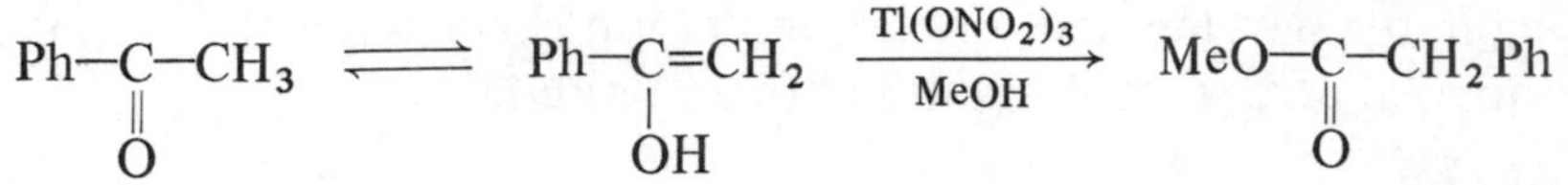

$$\underset{\underset{O}{\|}}{Ph-C-CH_3} \;\rightleftharpoons\; \underset{\underset{OH}{|}}{Ph-C=CH_2} \;\xrightarrow[MeOH]{Tl(ONO_2)_3}\; \underset{\underset{O}{\|}}{MeO-C-CH_2Ph}$$

In the case of alkynes, a keto carbonium ion is produced, and subsequent attack by the nucleophile can occur either at the carbonium centre or, provided an aryl migration can occur, at the adjacent carbonyl carbon.

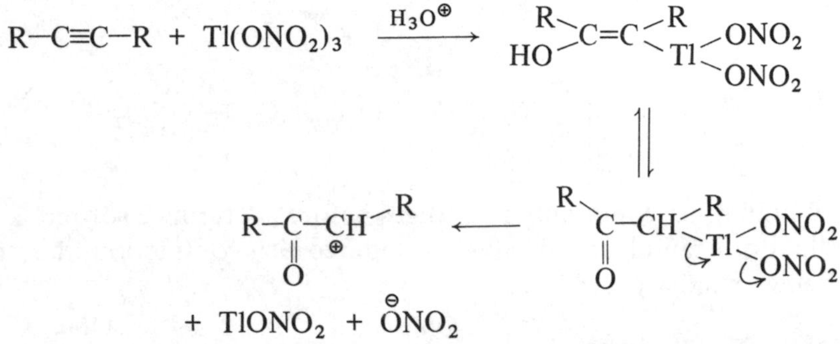

$$R-C\equiv C-R \;+\; Tl(ONO_2)_3 \;\xrightarrow{H_3O^{\oplus}}\; \underset{HO}{\overset{R}{>}}C=C\underset{Tl}{\overset{R}{<}}\underset{ONO_2}{\overset{ONO_2}{<}}$$

$$\underset{\underset{O}{\|}}{R-C-CH}\overset{R}{\underset{\oplus}{<}} \;\longleftarrow\; \underset{\underset{O}{\|}}{\overset{R}{>}C-CH}\underset{Tl}{\overset{R}{<}}\underset{ONO_2}{\overset{ONO_2}{<}}$$

$$+\; TlONO_2 \;+\; \overset{\ominus}{O}NO_2$$

Two possibilities for the further reaction of this intermediate carbonium ion are exemplified below. [232]

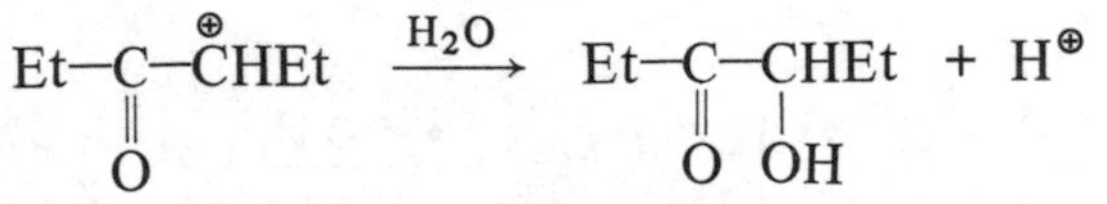

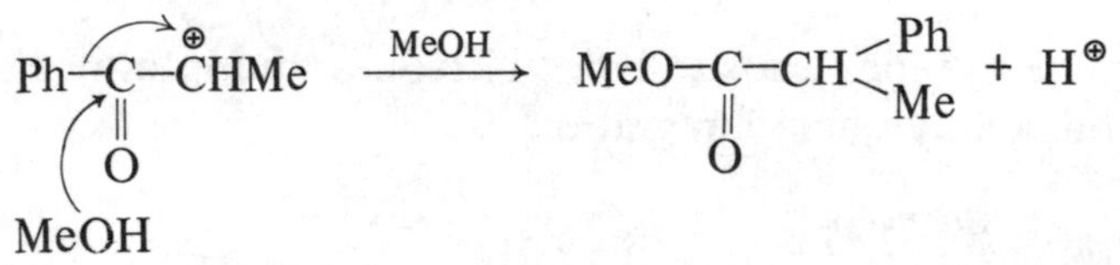

$$\underset{\underset{O}{\|}}{Et-\overset{\oplus}{C}-CHEt} \;\xrightarrow{H_2O}\; \underset{\underset{O}{\|}\;\;\underset{OH}{|}}{Et-C-CHEt} \;+\; H^{\oplus}$$

$$\underset{\underset{O}{\|}}{Ph-\overset{\oplus}{C}-CHMe} \;\xrightarrow{MeOH}\; \underset{\underset{O}{\|}}{MeO-C-CH}\underset{Me}{\overset{Ph}{<}} \;+\; H^{\oplus}$$

MeOH

Terminal alkynes undergo more extensive oxidation with two moles of thallium(III) nitrate to afford carboxylic acids having one carbon atom less than the starting alkyne.

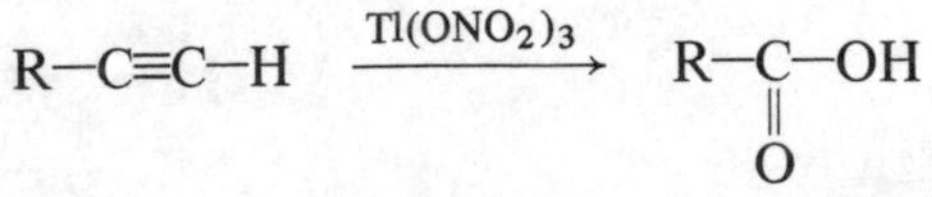

$$R-C\equiv C-H \;\xrightarrow{Tl(ONO_2)_3}\; \underset{\underset{O}{\|}}{R-C-OH}$$

Epoxides can be obtained from alkenes by oxidation with thallium(III) acetate. [233]

$$\underset{Me}{\overset{Me}{>}}C=CH_2 \; + \; Tl(OAc)_3 \; + \; H_2O \; \longrightarrow \; \underset{Me}{\overset{Me}{>}}\underset{\underset{OH}{|}}{C}-\underset{\underset{Tl(OAc)_2}{}}{CH_2} \; + \; HOAc$$

$$\underset{Me}{\overset{Me}{>}}\underset{\underset{O}{\bigtriangleup}}{C}-CH_2 \; + \; TlOAc \; + \; HOAc$$

## 7.6    Organic complexes of palladium

### 7.6.1    *Aromatic substitution*

Unlike aromatic mercuration and thallation, the direct reaction of aromatic hydrocarbons with palladium salts is an unsubstantiated process. However, it is postulated to occur in the palladium(II)-catalysed substitution of aromatic compounds by a wide variety of nucleophiles in the presence of a suitable oxidant, such as lead tetra-acetate.

$$\text{(benzene)} \; + \; X^{\ominus} \; + \; \text{oxidant} \; \xrightarrow{\text{Pd(II)}} \; \text{(benzene)}-X$$

Carbon–oxygen, carbon–nitrogen, carbon–sulphur and carbon–halogen bonds can be formed by the use of anions ($X^{\ominus}$) such as acetate, azide, thiocyanate and chloride. The role of the oxidant is not clear, but it could be required for the decomposition of an aryl palladium intermediate.

$$\text{(benzene)} \; + \; PdX_2 \; \xrightarrow{?} \; \text{(benzene)}-PdX \; \xrightarrow[X^{\ominus}]{\text{oxidant}} \; \text{(benzene)}-X \; + \; PdX_2$$

Alternatively, the oxidant may oxidize the palladium(II) to a palladium(IV) salt; the latter compounds are known to be capable of oxidizing arenes to the observed reaction products. [234]

### 7.6.2    *Benzylic acetoxylation*

Alkyl benzenes can be oxidized to benzylic acetates by palladium acetate in acetic acid.

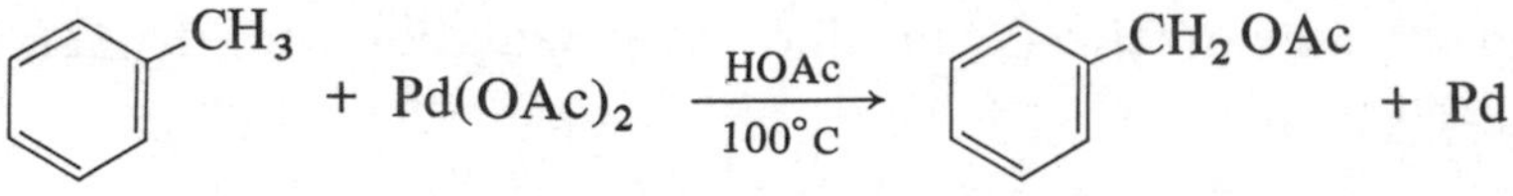

$$\text{(toluene)} + Pd(OAc)_2 \xrightarrow[100°C]{HOAc} \text{(benzyl acetate)} + Pd$$

Free-radical benzylic coupling is a competing process, but can be suppressed by the addition of excess potassium acetate. The acetoxylation reaction is enhanced by the presence of electron-donating substituents on the benzene ring and retarded by electron-withdrawing substituents. A posulated mechanism involves formation of an intermediate $\pi$-bonded aryl palladium species, which undergoes rearrangement to a benzylic $\sigma$-bonded palladium compound, which finally undergoes decomposition to the benzylic acetate. [235]

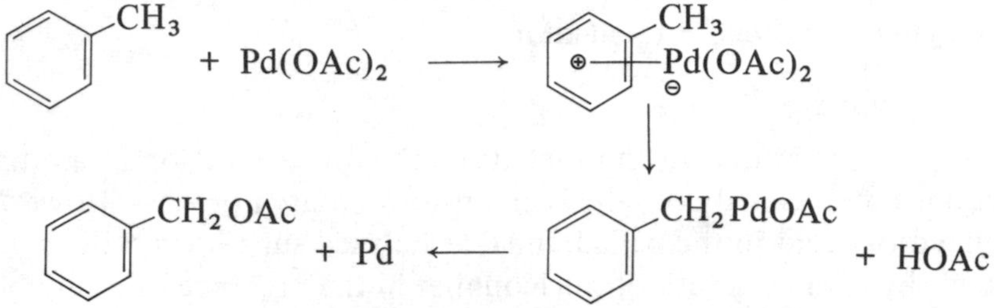

$$\text{(toluene)} + Pd(OAc)_2 \longrightarrow \text{(π-complex)}$$

$$\text{(benzyl acetate)} + Pd \longleftarrow \text{(CH}_2\text{PdOAc)} + HOAc$$

### 7.6.3  Reactions of nucleophiles with alkene complexes of palladium

Palladium salts readily form $\pi$-complexes with alkenes and these complexes are susceptible to attack by nucleophiles, either at the palladium atom or at an alkene carbon atom. Nucleophiles such as hydroxide ion, acetate ion, alkoxide ion and amines preferentially attack at carbon rather than palladium, to give a $\sigma$-bonded organopalladium compound with a new carbon–oxygen or carbon–nitrogen bond.

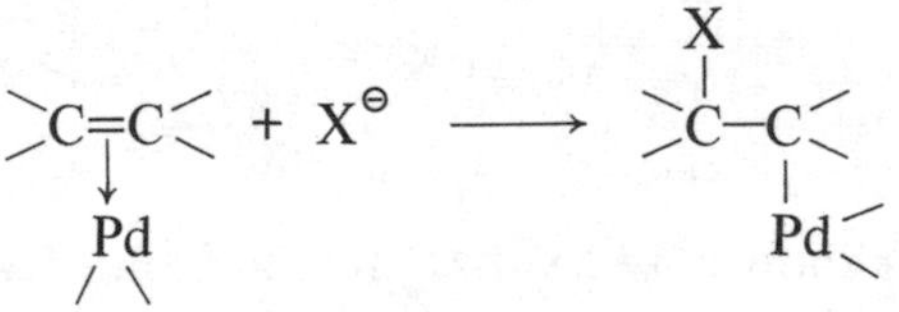

$$>C=C< \; + \; X^\ominus \longrightarrow \; >\underset{Pd<}{\overset{X}{C}}-C<$$

Such products can also be derived from the corresponding mercury or thallium organometallic compounds formed by oxymercuration or oxythallation.

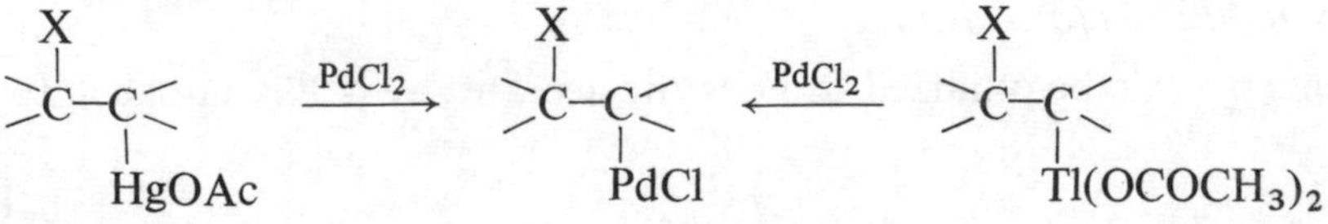

$$>\underset{HgOAc}{\overset{X}{C}}-C< \xrightarrow{PdCl_2} >\underset{PdCl}{\overset{X}{C}}-C< \xleftarrow{PdCl_2} >\underset{Tl(OCOCH_3)_2}{\overset{X}{C}}-C<$$

128

The carbon–palladium bond in these $\sigma$-bonded organopalladium compounds can be converted into a carbon–hydrogen bond in several ways, depending on the nature of X and the reaction conditions. The extensive work in this area is an outgrowth of the important discovery of the palladium-catalysed oxidation of ethene to acetaldehyde.

In 1894, Phillips showed that if ethene is passed into aqueous palladium chloride, the initially brown solution turns black due to precipitation of palladium metal. In this reaction an ethene–palladium complex is clearly involved. Kharasch in 1938 prepared $(PhCN)_2PdCl_2$ and on treating this with ethene in a non-aqueous medium obtained a binuclear complex.

This complex was found to decompose in water to yield palladium metal and acetaldehyde, and very significantly from the viewpoint of mechanism, when the complex is decomposed in deuterium oxide, the resultant acetaldehyde contains no deuterium. This means that the four hydrogen atoms initially present in the ethene molecule are still present in the acetaldehyde product, so that one hydrogen atom must migrate from one carbon to the other during the oxidative reaction.

A solution of palladium chloride in dilute hydrochloric acid contains the anions $[PdCl_4]^{2\ominus}$ and $[PdCl_3(OH)]^{2\ominus}$, both of which in the presence of ethene can exchange a chloride ion for an ethene unit, giving rise to the complex anions $[PdCl_3(C_2H_4)]^{\ominus}$ and $[PdCl_2(OH)(C_2H_4)]^{\ominus}$. In such an ethene–palladium complex there is donation of $\pi$-electron density from the carbon–carbon double bond into a vacant orbital of the palladium atom, so that the two carbons of the ethene unit can be regarded as acquiring a partial positive charge (incipient carbonium ion character – see Section 2.3). Slow hydrolysis of the dichlorohydroxy complex then results in insertion of the palladium atom and the co-ordinated hydroxy group across the alkene double bond, giving rise to a transient $\beta$-hydroxyethylpalladium(II) complex cation.

129

This latter species then undergoes rapid decomposition via a hydride ion shift from the β- to the α-carbon atom. This migration is accompanied by fission of the palladium–carbon σ-bond and reduction of the palladium(II) atom to palladium metal.

$$\left[ \begin{array}{c} Cl \\ Cl \end{array} \underset{Pd}{\overset{}{\diagdown}} \underset{OH_2}{\overset{CH_2-C-OH}{\diagup}} \overset{H}{\underset{H}{}} \right]^{\ominus} \xrightarrow{fast} 2Cl^{\ominus} + H_2O + Pd + CH_3\overset{\oplus}{C}\diagup^{OH}_{\diagdown H}$$

The liberated organic product is protonated acetaldehyde which loses a proton to the solvent.

$$CH_3-\underset{\oplus}{\overset{H}{\underset{|}{C}}}-OH + H_2O \longrightarrow CH_3-C\diagdown^{H}_{\diagdown O} + H_3O^{\oplus}$$

The industrially important Wacker process for the preparation of acetaldehyde from ethene requires only that the palladium metal be continuously reoxidized back to palladium(II), and this can be achieved by oxygen in the presence of a copper(II) salt. The following partial equations can be written, the overall process being the catalytic oxidation of ethene to acetaldehyde (ethanal). [236, 237]

$$CH_2{=}CH_2 + PdCl_4^{2\ominus} + H_2O \longrightarrow CH_3CHO + Pd + 2HCl + 2Cl^{\ominus}$$

$$2CuCl_2 + Pd + 2Cl^{\ominus} \longrightarrow Cu_2Cl_2 + PdCl_4^{2\ominus}$$

$$2HCl + \tfrac{1}{2}O_2 + Cu_2Cl_2 \longrightarrow 2CuCl_2 + H_2O$$

Sum:     $CH_2{=}CH_2 + \tfrac{1}{2}O_2 \longrightarrow CH_3CHO$

It has been suggested that the hydride transfer decomposition step takes place via an intermediate palladium hydrido complex, so that the hydrogen atom moves first from carbon to palladium and then from the metal to the other carbon atom (cf. Section 4.2.2).

$$\left[ \begin{array}{c} Cl \\ Cl \end{array} \underset{Pd}{\overset{}{\diagdown}} \underset{OH_2}{\overset{CH_2-CH}{\diagup}} \underset{H \ \ OH}{} \right]^{\ominus} \xrightarrow{-H_2O} \left[ \begin{array}{c} Cl \\ Cl \end{array} \underset{Pd}{\overset{H_2C}{\diagdown}} \underset{H}{\overset{CHOH}{\diagup}} \right]^{\ominus} \longrightarrow$$

$$2Cl^{\ominus} + Pd + CH_3-\underset{\oplus}{\overset{H}{\underset{|}{C}}}-OH$$

Further support for a mechanism involving hydroxyl ion attack on an incipient carbonium ion attached to the palladium atom is found in the observation that when an unsymmetrical alkene is oxidized with air, palladium chloride and copper(II) chloride, the carbon–oxygen bond is formed at that carbon atom where nucleophilic attack would be predicted according to Markovnikov's rule (attack on the more stable carbonium ion). Thus with a terminal alkene, $RCH=CH_2$, the product is the methyl ketone, $RCOCH_3$.

$$PdCl_4^{2\ominus} + RCH=CH_2 \rightleftharpoons [PdCl_3(RCH=CH_2)]^{\ominus} + Cl^{\ominus}$$

$$[PdCl_3(RCH=CH_2)]^{\ominus} + H_2O \rightleftharpoons [PdCl_2(OH)(RCH=CH_2)]^{\ominus} + HCl$$

$$\left[ \begin{array}{c} Cl \quad H_2C \\ \phantom{Cl}\diagdown Pd \diagup \phantom{} CHR \\ Cl \diagup \phantom{Pd} \diagdown OH \end{array} \right]^{\ominus}$$

$\downarrow H_2O$

$$\left[ \begin{array}{c} Cl \diagdown \phantom{Pd} CH_2-CHR \\ \phantom{Cl}Pd \phantom{-} | \\ Cl \diagup \phantom{Pd} OH_2 \quad OH \end{array} \right]^{\ominus} \quad \text{and not} \quad \left[ \begin{array}{c} R \\ | \\ Cl \diagdown \phantom{Pd} CH-CH_2OH \\ \phantom{Cl}Pd \\ Cl \diagup \phantom{Pd} OH_2 \end{array} \right]^{\ominus}$$

$\downarrow$

$$2Cl^{\ominus} + Pd + H_3O^{\oplus} + CH_3COR$$

The reaction of certain branched alkenes affords unsaturated carbonyl compounds via an intermediate $\pi$-allylic palladium complex. This complex is formed by the elimination of hydrogen chloride from the initial $\pi$-alkene complex, and then undergoes oxidation by a further molecule of palladium chloride. In this way isobutene (2-methylpropene) can be oxidized to 2-methylacrolein (2-methylpropenal).

131

$$CH_2=C{<}{\overset{CH_3}{CH_3}} \ + \ [PdCl_3(OH)]^{2\ominus} \ \longrightarrow \ \left[\begin{array}{c} {}_{H_2C}{\diagdown}{\underset{C}{\parallel}}{\diagup}{}^{CH_3} \\ {}_{Cl}{\diagdown}{}_{Pd}{\diagup}{}^{C}{\diagdown}{}_{CH_3} \\ {}_{HO}{\diagup}{\phantom{Pd}}{\diagdown}{}_{Cl} \end{array}\right]^{\ominus} + \ Cl^{\ominus}$$

$$\downarrow \ {-HCl}$$

$$CH_2=C{<}{\overset{CH_3}{CHO}} \quad \xleftarrow[H_2O]{PdCl_2} \quad \left[\begin{array}{c} {}_{Cl}{\diagdown}{\phantom{Pd}}{\diagup}{}^{CH_2}{\diagdown} \\ {}_{\phantom{Cl}}{Pd}{\phantom{aa}}{\vdots}{\phantom{a}}C{-}CH_3 \\ {}_{HO}{\diagup}{\phantom{Pd}}{\diagdown}{}_{CH_2}{\diagup} \end{array}\right]^{\ominus}$$

While no reaction other than complex formation occurs with ethene and palladium chloride in acetic acid at greater than 80% concentration, the addition of sodium acetate leads to the formation of palladium metal and vinyl acetate as a consequence of nucleophilic attack by acetate ion on the coordinated ethene. [238] This reaction has been developed as a commercial route to vinyl acetate and hence poly(vinyl acetate) and poly(vinyl alcohol).

$$-\overset{|}{\underset{|}{Pd}}{\leftarrow}\overset{CH_2}{\underset{CH_2}{\parallel}} \ + \ AcO^{\ominus} \ \longrightarrow \ -\overset{|}{\underset{|}{Pd}}-CH_2-CH_2-OAc$$

$$\downarrow OAc^{\ominus}$$

$$-\overset{|}{\underset{|}{Pd}}^{\ominus}+ \ CH_2=CHOAc \ + \ HOAc$$

Significant quantities of 1,1-diacetoxyethane are also formed, and this result is consistent with a hydride ion migration during decomposition of the β-acetoxyethylpalladium complex, with liberation of a carbonium ion which can then either lose a proton to give vinyl acetate or capture a second acetoxy anion to give 1,1-diacetoxyethane.

$$-\overset{|}{\underset{|}{Pd}}{-}CH_2{-}\overset{\overset{H}{|}}{CH}{-}OAc \ \longrightarrow \ -\overset{|}{\underset{|}{Pd}}{}^{\ominus} + \ CH_3{-}\overset{\oplus}{CH}OAc$$

$${}^{\ominus}OAc \ attack \ on \ carbon \diagup \qquad \qquad \diagdown {}^{\ominus}OAc \ attack \ on \ hydrogen$$

$$CH_3CH(OAc)_2 \qquad CH_2=CHOAc \ + \ HOAc$$

Alternatively, the vinyl acetate may arise by direct elimination of a palladium(II) hydride or by intramolecular attack of the carbonyl oxygen on the methylene group attached to palladium. In the latter case, the cyclic acetoxonium ion would then yield vinyl acetate as a consequence of further attack by acetate ion on any one of the four hydrogen atoms.

Some support for a mechanism of this kind is found in the conversion of norbornene into *exo*-2-chloro-*syn*-7-acetoxynorbornene by palladium chloride in acetic acid, which can be written: [239]

The reaction of higher alkenes with palladium acetate has also been studied. In general a terminal alkene affords the corresponding vinyl acetate as the main product whereas other alkenes are predominantly converted into allylic acetates.

$$RCH_2CH=CH_2 \xrightarrow{Pd(OAc)_2}$$

$$RCH_2-\underset{\underset{OAc}{|}}{CH}-CH_2PdOAc \xrightarrow{-HPdOAc} RCH_2-\underset{\underset{OAc}{|}}{C}=CH_2$$

$$RCH_2CH=CHCH_3 \xrightarrow{Pd(OAc)_2}$$

$$RCH_2-\underset{\underset{OAc}{|}}{CH}-\underset{\underset{PdOAc}{|}}{CH}CH_3 \xrightarrow{-HPdOAc} RCH_2\underset{\underset{OAc}{|}}{CH}-CH=CH_2$$

Acetoxy palladation is followed by elimination of acetoxyhydridopalladium and hydride migration from one carbon atom to another is not observed. Closely related reactions, the study of which has yielded much mechanistic information, are the vinylic and allylic exchange reactions represented by the following equations in which X or Y = OCOR, Cl, OR, $NR_2$ etc.). [240, 241]

$$CH_2=CHX + Y^\ominus \underset{}{\overset{Pd(II)}{\rightleftharpoons}} CH_2=CHY + X^\ominus$$

$$CH_2=CH-CH_2X + Y^\ominus \underset{}{\overset{Pd(II)}{\rightleftharpoons}} CH_2=CH-CH_2Y + X^\ominus$$

These reactions are non-oxidative in nature, a palladium(II) dimer complex is the reactive species, and the mechanism involves *trans* addition to the double bond followed by *trans* elimination (see Section 2.7). For the vinyl ester exchange in acetic acid containing palladium chloride and lithium chloride, i.e.,

$$CH_2=CH-OCOR + HOAc \underset{}{\overset{Pd(II)}{\rightleftharpoons}} CH_2=CH-OCOCH_3 + HOCOR$$

the following mechanism can be written:

$$\left[ \begin{array}{c} \text{Cl}\diagdown\text{Pd}\diagdown\text{Cl}\diagup\text{Pd}\diagup\substack{\text{H}_2\text{C}\\ \text{CHOCOR}} \\ \text{Cl}\diagup\quad\text{Cl}\quad\text{Cl} \end{array} \right]^{\ominus} \rightleftharpoons$$

$$-\overset{\ominus}{\text{O}}\text{Ac}$$

$$\left[ \begin{array}{c} \text{Cl}\diagdown\text{Pd}\diagdown\text{Cl}\diagup\text{Pd}\diagup\substack{\text{CH}_2-\text{CH}\diagdown^{\text{OCOR}}\\ \text{OAc}} \\ \text{Cl}\diagup\quad\text{Cl}\quad\text{Cl} \end{array} \right]^{2\ominus} \rightleftharpoons$$

$$\left[ \begin{array}{c} \text{Cl}\diagdown\text{Pd}\diagdown\text{Cl}\diagup\text{Pd}\diagup\substack{\text{H}_2\text{C}\\ \text{CHOAc}} \\ \text{Cl}\diagup\quad\text{Cl}\quad\text{Cl} \end{array} \right]^{\ominus} + \text{RCO}_2^{\ominus}$$

In similar fashion, vinyl and allyl chlorides can be converted into the corresponding acetates, ethers and amines. [242, 243]

Acetals and to a lesser extent vinyl ethers can be prepared by reaction of alcohols with alkene–palladium complexes. [244]

$$CH_2\text{=}CH_2 \xrightarrow[\text{Pd(II)}]{\text{ROH}} CH_2\text{=}CHOR + CH_3CH(OR)_2$$

In the presence of water the acetal is hydrolysed to the corresponding carbonyl compound.

The nucleophilic addition of primary amines to an alkene–palladium complex affords a σ-bonded metal complex containing a new carbon–nitrogen bond. Elimination of the metal hydride (see Section 2.7) gives rise to an imine, while hydrogenation of the complex yields a secondary amine. [9, 245]

$$RCH\text{=}CH_2 \xrightarrow[\text{Pd(II)}]{\text{R'NH}_2} \underset{\text{R'NH}}{RCH-CH_2-Pd-} \xrightarrow{-\,HPd-} \underset{\text{R'NH}}{R-C\text{=}CH_2}$$

$$\downarrow \text{H}_2/\text{Ni or NaBH}_4$$

$$\underset{\text{R'}\diagdown^{\text{NH}}}{R\diagdown_{\text{CH}}\diagup\text{CH}_3}$$

$$\underset{\text{R'}\diagdown^{\text{N}}}{R\diagdown_{\text{C}}\diagup\text{CH}_3}$$

7.7    Carbamoyl complexes of transition metals – formation of carbon-nitrogen bonds

Primary amines react with carbon monoxide in the presence of palladium chloride to yield isocyanates, and the reaction is thought to involve a 'carbonyl insertion' step (cf. Chapter 5) involving migration of the nitrogen atom from palladium to carbon. [246]

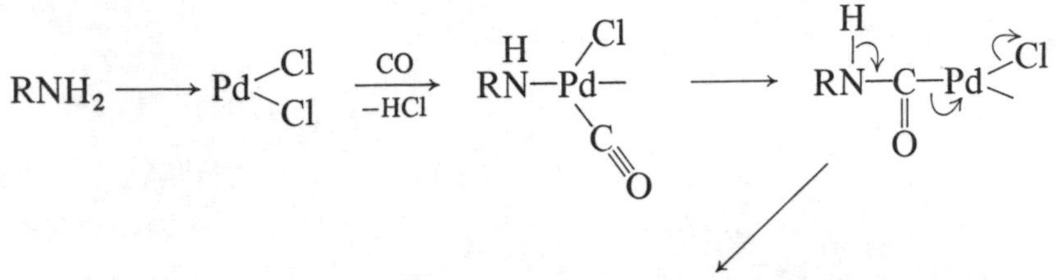

$$RNH_2 \longrightarrow Pd\overset{Cl}{\underset{Cl}{<}} \xrightarrow[-HCl]{CO} \overset{H}{\underset{\overset{|}{C}=O}{RN-Pd-}}\overset{Cl}{/} \longrightarrow \overset{H}{\underset{\overset{||}{O}}{RN-C-Pd}}\overset{Cl}{<}$$

$$H^{\oplus} + R-N=C=O + Pd + Cl^{\ominus}$$

Such metal-catalysed reactions of amines with carbon monoxide to form organic isocyanates, formamides, and ureas are well known in the industrial field, and it seems likely that carbamoyl metal complexes are intermediates in these reactions. [247] Carbamoyl complexes have recently been prepared and isolated and can be made to yield isocyanates and ureas under appropriate conditions.

Direct reaction of the cyclopentadienyltetracarbonyltungsten cation with a secondary amine yields the neutral cyclopentadienyltricarbonyl carbamoyl tungsten compound.

$$(C_5H_5)W(CO)_4^{\oplus} + 2HNRR' \rightleftharpoons (C_5H_5)W(CO)_3C\overset{O}{\underset{NRR'}{<}} + H_2\overset{\oplus}{N}RR'$$

Many similar complexes have been prepared using appropriate carbonyl derivatives of other metals such as platinum, ruthenium, iron, manganese, molybdenum and rhenium, and a general equation can be written

$$L_nM-C\equiv\overset{\oplus}{O} + 2HNRR' \rightleftharpoons L_nM-C\overset{O}{\underset{NRR'}{<}} + H_2\overset{\oplus}{N}RR'$$

The above carbamoyl tungsten complexes react with tertiary amines to give isocyanates and with primary amines to yield ureas.

$$(C_5H_5)W(CO)_3C{\overset{O}{\underset{NHCH_3}{}}} + Et_3N \rightleftharpoons$$

$$(C_5H_5)\overset{\ominus}{W}(CO)_3 + Et_3\overset{\oplus}{N}H + CH_3NCO$$

$$(C_5H_5)W(CO)_3C{\overset{O}{\underset{NHCH_3}{}}} + 2CH_3NH_2 \longrightarrow$$

$$(C_5H_5)\overset{\ominus}{W}(CO)_3 + CH_3\overset{\oplus}{N}H_3 + CH_3NH\underset{O}{\overset{||}{C}}NHCH_3$$

The analogous molybdenum complex reacts similarly with tertiary amines, but the more stable $(C_5H_5)Fe(CO)_2CONHCH_3$ does not. However this latter substance gives rise to methylisocyanate on treatment with bromine in the following series of reactions:

$$2(C_5H_5)Fe(CO)_2CONHCH_3 + 2Br_2 \longrightarrow$$

$$2(C_5H_5)Fe(CO)_2Br + 2BrCONHCH_3$$

$$2BrCONHCH_3 \longrightarrow 2HBr + 2CH_3NCO$$

$$(C_5H_5)Fe(CO)_2CONHCH_3 + 2HBr \longrightarrow$$

$$[(C_5H_5)Fe(CO)_3]^{\oplus}Br^{\ominus} + CH_3\overset{\oplus}{N}H_3\ Br^{\ominus}$$

Sum: $3(C_5H_5)Fe(CO)_2CONHCH_3 + 2Br_2 \longrightarrow$

$2(C_5H_5)Fe(CO)_2Br + [(C_5H_5)Fe(CO)_3]^{\oplus}Br^{\ominus} + CH_3\overset{\oplus}{N}H_3\ Br^{\ominus} + 2CH_3NCO$

These and other reactions of metal carbamoyl complexes and the related alkoxycarbonyl complexes, $L_nM—C{\overset{O}{\underset{OR}{}}}$ , have been reviewed. [248]

Lithium dimethylamide reacts with tetracarbonylnickel to give a tricarbonyl-carbamoylnickel anion which on subsequent reaction with a reactive alkyl or acyl halide yields the corresponding carboxylic acid dimethylamide. [249] As explained in Section 2.9, the final step in each synthesis is probably a solvent-induced reductive elimination.

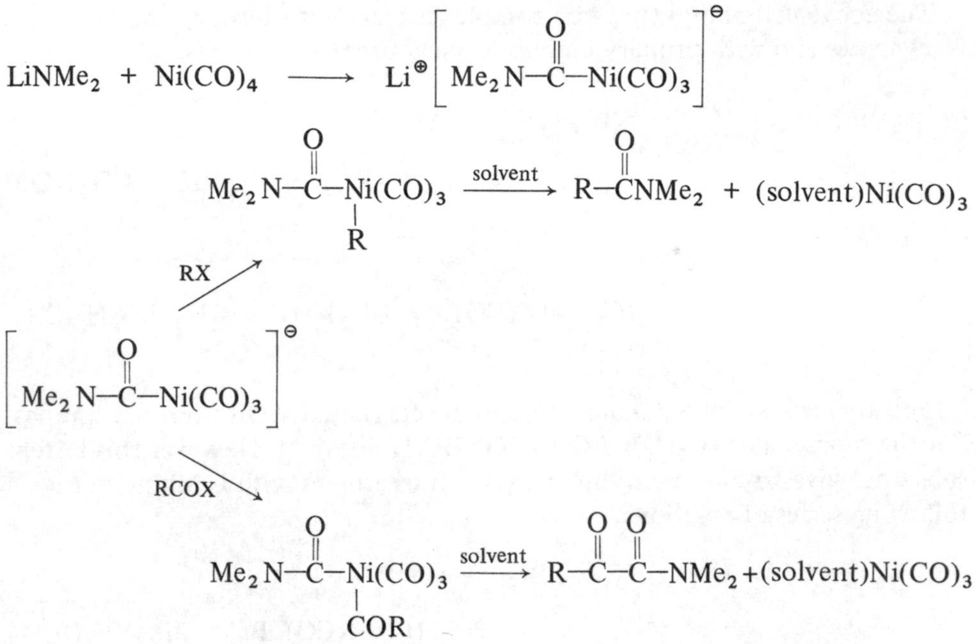

$$LiNMe_2 + Ni(CO)_4 \longrightarrow Li^{\oplus} \left[ Me_2N-\overset{\overset{\displaystyle O}{\|}}{C}-Ni(CO)_3 \right]^{\ominus}$$

$$Me_2N-\overset{\overset{\displaystyle O}{\|}}{\underset{\underset{\displaystyle R}{|}}{C}}-Ni(CO)_3 \xrightarrow{\text{solvent}} R-\overset{\overset{\displaystyle O}{\|}}{C}NMe_2 + (\text{solvent})Ni(CO)_3$$

RX ↗

$$\left[ Me_2N-\overset{\overset{\displaystyle O}{\|}}{C}-Ni(CO)_3 \right]^{\ominus}$$

RCOX ↘

$$Me_2N-\overset{\overset{\displaystyle O}{\|}}{\underset{\underset{\displaystyle COR}{|}}{C}}-Ni(CO)_3 \xrightarrow{\text{solvent}} R-\overset{\overset{\displaystyle O}{\|}}{C}-\overset{\overset{\displaystyle O}{\|}}{C}-NMe_2 + (\text{solvent})Ni(CO)_3$$

## 7.8    Organorhodium complexes—formation of carbon–halogen bonds

The Wilkinson hydrogenation catalyst tris(triphenylphosphine)chlororhodium (see Section 6.5) effects decarbonylation of aroyl chlorides to aryl chlorides [250] (cf. Section 5.2).

$$ArCOCl + (Ph_3P)_3RhCl \longrightarrow ArCl + (Ph_3P)_2Rh(CO)Cl + Ph_3P$$

Phenylacetyl chloride yields benzyl chloride, while benzoyl bromide yields bromobenzene. Similar reactions occur with arene sulphonyl chlorides.

$$ArSO_2Cl \xrightarrow[\text{or } (Ph_3P)_4Pt]{(Ph_3P)_3RhCl} ArCl$$

These reactions can all be interpreted as an oxidative addition to the rhodium atom followed by a reverse migration reaction (cf. Section 2.6) and a solvent-induced reductive elimination (cf. Section 2.5).

# References

[1]    Wilke, G. (1963), *Angew. Chem. Internat. Edit.*, **2**, 105.
[2]    Brown, H. C. and Negishi, E. (1967), *J. Am. Chem. Soc.*, **89**, 5478.
[3]    Stetter, H. (1963), in Foerst, W., Ed., 'Newer Methods of Preparative Organic Chemistry', Vol. 2, Academic Press, New York, pp. 51–99.
[4]    Chatt, J. (1951), *Chem. Rev.*, **48**, 7.
[5]    Kitching, W. (1968), *Organometal. Chem. Rev.*, **3**, 35, 61.
[6]    Traylor, T. G. (1969), *Accts Chem. Res.*, **2**, 152.
[7]    Smidt, J., Hafner, W., Jira, R., Sedlomeir, J., Sieber, R., Ruttinger, R. and Kojer, H. (1959), *Angew. Chem.*, **71**, 176.
[8]    Tsuji, J. (1969), *Accts Chem. Res.*, **2**, 144.
[9]    Stille, J. K. and Fox, D. B. (1970), *J. Am. Chem. Soc.*, **92**, 1274.
[10]   Tsuji, J. and Takahashi, H. (1965), *J. Am. Chem. Soc.*, **87**, 3275.
[11]   Takahashi, H. and Tsuji, J. (1968), *J. Am. Chem. Soc.*, **90**, 2387.
[12]   Johnson, B. F. G., Lewis, J. and Subramaniam, M. S. (1968), *J. Chem. Soc. (A)*, 1993.
[13]   Casey, C. P. and Burkhardt, T. J. (1972), *J. Am. Chem. Soc.*, **94**, 6543.
[14]   Moser, W. R. (1969), *J. Am. Chem. Soc.*, **91**, 1135, 1141.
[15]   Bacon, R. G. R. and Hill, H. A. O. (1965), *Quart. Rev.*, **19**, 95.
[16]   Eglinton, G. and McCrae, W. (1963), *Adv. Org. Chem.*, **4**, 225.
[17]   Kharasch, M. S. and Reinmuth, O. (1954), 'Grignard Reactions of Non-Metallic Substances', Prentice-Hall Inc., Englewood Cliffs, N.J.
[18]   Kaufmann, T. and Sahm, W. (1967), *Angew. Chem. Internat. Edit.*, **6**, 85.
[19]   Corriu, R. J. P. and Masse, J. P. (1972), *J.C.S. Chem. Comm.*, 144.
[20]   Tamao, K., Sumitani, K. and Kumada, M. (1972), *J. Am. Chem. Soc.*, **94**, 4374.
[21]   Davies, A. G. and Roberts, B. P. (1972), *Accts Chem. Res.*, **5**, 387.
[22]   Bauld, N. L. (1962), *Tetrahedron Lett.*, 859.
[23]   McKillop, A., Elsom, L. F. and Taylor, E. C. (1968), *J. Am. Chem. Soc.*, **90**, 2423; (1970), *Tetrahedron*, **26**, 4041.
[24]   Taylor, E. C., Kienzle, F. and McKillop, A. (1970), *J. Am. Chem. Soc.*, **92**, 6088.
[25]   Uemura, S., Ikeda, Y. and Ichikawa, K. (1971), *Chem. Comm.*, 390.
[26]   Tolman, C. A. (1972), *Chem. Soc. Rev.*, **1**, 337; Schwartz, J., Hart, D. W. and Holden, J. L. (1972), *J. Am. Chem. Soc.*, **94**, 9269; Schwartz, J. and Cannon, J. B. (1972), *J. Am. Chem. Soc.*, **94**, 6226.
[27]   Wender, I., Levine, R. and Orchin, M. (1949), *J. Am. Chem. Soc.*, **71**, 4160.
[28]   Deacon, G. B. and Felder, P. W. (1969), *Aust. J. Chem.*, **22**, 549.
[29]   Whitesides, G. M. and Boschetto, D. J. (1971), *J. Am. Chem. Soc.*, **93**, 1529.

[30] Brown, H. C. (1969), *Accts Chem. Res.*, **2**, 65.

[31] Heck, R. F. (1968), *J. Am. Chem. Soc.*, **90**, 5518 and following papers.

[32] Dey, K., Eaborn, C. and Walton, D. R. M. (1970/1971), *Organometal. Chem. Syn.*, **1**, 151.

[33] Ziegler, K. and Gellert, H. G. (1950), *Liebigs Ann.*, **567**, 179.

[34] Finnegan, R. A. and Kutta, H. W. (1965), *J. Org. Chem.*, **30**, 4138.

[35] Rodeheaver, G. T. and Hunt, D. F. (1971), *Chem. Comm.*, 818.

[36] Kretchmer, R. A., Conrad, R. A. and Mihelich, E. D. (1973), *J. Org. Chem.*, **38**, 1251.

[37] Seyferth, D., Burlitch, J. M. and Heeren, J. K. (1962), *J. Org. Chem.*, **27**, 1491.

[38] Seyferth, D. and Burlitch, J. M. (1964), *J. Am. Chem. Soc.*, **86**, 2730.

[39] Seyferth, D. (1972), *Accts Chem. Res.*, **5**, 65.

[40] Clark, H. C. and Willis, C. J. (1960), *J. Am. Chem. Soc.*, **82**, 1888.

[41] Seyferth, D., Mui, J. Y. P., Gordon, M. E. and Burlitch, J. M. (1965), *J. Am. Chem. Soc.*, **87**, 681.

[42] Simmons, H. E. and Smith, R. D. (1959), *J. Am. Chem. Soc.*, **81**, 4256.

[43] Blanchard, E. P. and Simmons, H. E. (1964), *J. Am. Chem. Soc.*, **86**, 1337.

[44] Simmons, H. E., Blanchard, E. P. and Smith, R. D. (1964), *J. Am. Chem. Soc.*, **86**, 1347.

[45] Rawson, R. J. and Harrison, I. T. (1970), *J. Org. Chem.*, **35**, 2057.

[46] Furukawa, J., Kawabata, N. and Nishimura, J. (1968), *Tetrahedron*, **24**, 53; Nishimura, J. and Furukawa, J. (1971), *Chem. Comm.*, 1375.

[47] Hoberg, H. (1962), *Liebigs Ann.*, **656**, 1.

[48] Cardin, D. J., Cetinkaya, B., Doyle, M. J. and Lappert, M. F. (1973), *Chem. Soc. Rev.*, **2**, 99.

[49] Fischer, E. O., Heckl, B., Dotz, K. H., Müller, J. and Werner, H. (1969), *J. Organometal. Chem.*, **16**, P29.

[50] Fischer, E. O. and Dötz, K. H. (1972), *J. Organometal. Chem.*, **36**, C4.

[51] Cardin, D. J., Doyle, M. J. and Lappert, M. F. (1972), *J.C.S. Chem. Comm.*, 927.

[52] Corey, E. J. and Hegedus, L. S. (1969), *J. Am. Chem. Soc.*, **91**, 4926.

[53] Birch, A. J., Cross, P. E., Lewis, J., White, D. A. and Wild, S. B. (1968), *J. Chem. Soc. (A)*, 332.

[54] Birch, A. J. and Haas, M. (1971), *J. Chem. Soc. (C)*, 2465.

[55] Nicholls, B. and Whiting, M. C. (1959), *J. Chem. Soc.*, 551.

[56] Bolton, E. S., Knox, G. R. and Robertson, C. G. (1969), *Chem. Comm.*, 664.

[57] Johnson, B. F. G., Lewis, J., Parkins, A. W. and Randall, G. L. P. (1969), *Chem. Comm.*, 595.

[58] Nicholas, K. M. and Pettit, R. (1971), *Tetrahedron Lett.*, 3475.

[59] Seyferth, D. and Wehman, A. T. (1970), *J. Am. Chem. Soc.*, **92**, 5520.

[60] Arthur, P., England, D. C., Pratt, B. C. and Whitman, G. M. (1954), *J. Am. Chem. Soc.*, **76**, 5364.

[61] Pettit, R. and Emerson, G. F. (1964), *Adv. Organometal. Chem.*, **1**, 1; Maitlis, P. M. (1966), *Adv. Organometal. Chem.*, **4**, 95.

[62] Emerson, G. F., Watts, L. and Pettit, R. (1965), *J. Am. Chem. Soc.*, **87**, 131.

[63] Roth, W. and Meier, J. D. (1967), *Tetrahedron Lett.*, 2053.

[64] Emerson, G. F., Ehrlich, K., Giering, W. P. and Lautebur, P. C. (1966), *J. Am. Chem. Soc.*, **88**, 3172.

[65] Landesberg, J. M. and Sieczkowski, J. (1969), *J. Am. Chem. Soc.*, **91**, 2120.

[66] Seebach, D. (1969), *Synthesis*, **1**, 17.

[67] Evans, E. A. (1956), *J. Chem. Soc.*, 4691; Izzo, P. T. and Safir, S. R. (1959), *J. Org. Chem.*, **24**, 701.

[68]    Ryang, M. and Tsutsumi, S. (1962), *Bull. Chem. Soc. Japan*, **35**, 1121; Ryang, M., Sawa, Y., Hasimoto, T. and Tsutsumi, S. (1964), *Bull. Chem. Soc. Japan*, **37**, 1704

[69]    Corey, E. J. and Posner, G. H. (1967), *J. Am. Chem. Soc.*, **89**, 3911.

[70]    Dubois, J. E., Lion, C. and Moulineau, C. (1971), *Tetrahedron Lett.*, 177.

[71]    Humphrey, S. A., Herrmann, J. L. and Schlessinger, R. H. (1971), *Chem. Comm.*, 1244.

[72]    Posner, G. H., Whitten, C. E. and McFarland, P. E. (1972), *J. Am. Chem. Soc.*, **94**, 5106.

[73]    Herr, R. W., Wieland, D. M. and Johnson, C. R. (1970), *J. Am. Chem. Soc.*, **92**, 3813.

[74]    Frejaville, C. and Jullien, R. (1971), *Tetrahedron Lett.*, 2039; Casey, C. P. and Boggs, R. A. (1971), *Tetrahedron Lett.*, 2455.

[75]    House, H. O. and Fischer, W. F. (1968), *J. Org. Chem.*, **33**, 949.

[76]    Corey, E. J. and Kuwajima, I. (1970), *J. Am. Chem. Soc.*, **92**, 395.

[77]    Fieser, L. M. and Fieser, M. (1967), 'Reagents for Organic Synthesis', Wiley-Interscience, Vol. 1, p. 415.

[78]    Birch, A. J. and Robinson, R. (1943), *J. Chem. Soc.*, 501.

[79]    Kharasch, M. S. and Tawney, P. O. (1941), *J. Am. Chem. Soc.*, **63**, 2308.

[80]    Church, R. F., Ireland, R. E. and Shridhar, D. R. (1962), *J. Org. Chem.*, **27**, 707.

[81]    House, H. O., Latham, R. A. and Slater, C. D. (1966), *J. Org. Chem.*, **31**, 2667.

[82]    Mole, T. and Jeffery, E. A. (1972), 'Organoaluminium Compounds', Elsevier, Amsterdam.

[83]    Meisters, A. and Mole, T. (1972), *J.C.S. Chem. Comm.*, 595.

[84]    Nagata, W. and Yoshioka, M. (1966), *Tetrahedron Lett.*, 1913.

[85]    Nagata, W., Yoshioka, M. and Murakami, M. (1972), *J. Am. Chem. Soc.*, **94**, 4644, 4654.

[86]    Nagata, W., Yoshioka, M. and Terasawa, T. (1972), *J. Am. Chem. Soc.*, **94**, 4672.

[87]    Zweifel, G. and Steele, R. B. (1967), *J. Am. Chem. Soc.*, **89**, 2754.

[88]    Zweifel, G., Snow, J. T. and Whitney, C. C. (1968), *J. Am. Chem. Soc.*, **90**, 7139.

[89]    Zweifel, G. and Steele, R. B. (1967), *J. Am. Chem. Soc.*, **89**, 5085.

[90]    Brown, H. C. and Rogic, M. M. (1972), *Organometal. Chem. Syn.*, **1**, 305.

[91]    Corey, E. J. and Hamanaka, E. (1964), *J. Am. Chem. Soc.*, **86**, 1641; (1967), **89**, 2758; Corey, E. J. and Semmelhack, M. F. (1966), *Tetrahedron Lett.*, 6237; Corey, E. J. and Wat, E. K. W. (1967), *J. Am. Chem. Soc.*, **89**, 2757.

[92]    Semmelhack, M. F., Helquist, P. M. and Gorzynski, J. D. (1972), *J. Am. Chem. Soc.*, **94**, 9234.

[93]    Kunts, A. I., Beletskaya, I. P., Savchenko, L. A. and Reutov, O. A. (1969), *J. Organometal. Chem.*, **17**, P21.

[94]    Gilman, H., Wooley, B. L. and Wright, G. F. (1933), *J. Am. Chem. Soc.*, **55**, 2609; Gilman, H. and Nelson, J. (1936), *Rec. trav. Chim.*, **55**, 518; Chute, W. J., Orchard, W. M. and Wright, G. F. (1941), *J. Org. Chem.*, **6**, 157.

[95]    Noltes, J. G., Verbeek, F. and Creemers, H. M. J. C. (1970/1971), *Organometal. Chem. Syn.*, **1**, 57.

[96]    Vaughan, W. R. and Knoess, H. P. (1970), *J. Org. Chem.*, **35**, 2394; Taylor, E. C., Hawks, G. H. and McKillop, A. (1968), *J. Am. Chem. Soc.*, **90**, 2421.

[97]    Finnegan, R. A. (1963), *Tetrahedron Lett.*, 851.

[98]    Zakharkin, L. I., Okhlobystin, O. Y. and Strunin, B. N. (1965), *J. Organometal. Chem.*, **4**, 349.

[99]    Jerkunica, J. M. and Traylor, T. G. (1971), *J. Am. Chem. Soc.*, **93**, 6278.

[100]   Scouten, C. G., Barton, F. E., Burgess, J. R., Story, P. R. and Garst, J. F. (1969), *Chem. Comm.*, 78.

[101]   Jarvie, A. W. P. (1970), *Organometal. Chem. Rev. Sect. A.*, **6**, 153.

[102]   Cainelli, G., Berlini, F., Grasselli, P. and Zubiani, G. (1967), *Tetrahedron Lett.*, 5153.
[103]   Brown, H. C. and Zweifel, G. (1959), *J. Am. Chem. Soc.*, **81**, 1512; (1960), **82**, 3222, 3223; (1961), **83**, 1241, 3834.
[104]   Wilke, G. and Müller, H. (1956), *Chem. Ber.*, **89**, 444; Wilke, G. and Müller, H. (1960), *Liebigs Ann.*, **629**, 222.
[105]   Zweifel, G., Fisher, R. P., Snow, J. T. and Whitney, C. C. (1971), *J. Am. Chem. Soc.*, **93**, 6309.
[106]   Coffey, C. E. (1961), *J. Am. Chem. Soc.*, **83**, 1623.
[107]   Cassar, L., Eaton, P. E. and Halpern, J. (1970), *J. Am. Chem. Soc.*, **92**, 3515.
[108]   Katz, T. J. and Cerefice, S. A. (1971), *J. Am. Chem. Soc.*, **93**, 1049.
[109]   Gassman, P. G. and Atkins, T. J. (1971), *J. Am. Chem. Soc.*, **93**, 4597; Gassman, P. G. and Nakai, T. (1971), *J. Am. Chem. Soc.*, **93**, 5897; Gassman, P. G. and Williams, F. J. (1972), *Chem. Comm.*, 80.
[110]   Gassman, P. G., Atkins, T. J. and Lumb, J. T. (1971), *Tetrahedron Lett.*, 1643.
[111]   Sakai, M., Yamaguchi, H., Westberg, H. H. and Masamune, S. (1971), *J. Am. Chem. Soc.*, **93**, 1043.
[112]   Paquette, L. A., Wilson, S. E. and Henzel, R. P. (1971), *J. Am. Chem. Soc.*, **93**, 1288; Paquette, L. A. and Wilson, S. E. (1971), *J. Am. Chem. Soc.*, **93**, 5934.
[113]   Gassman, P. G. and Atkins, T. J. (1971), *J. Am. Chem. Soc.*, **93**, 1042.
[114]   Orchin, M. (1966), *Adv. Catalysis*, **16**, 1.
[115]   Harrod, J. F. and Chalk, A. J. (1964), *J. Am. Chem. Soc.*, **86**, 1776; Hartley, F. R. (1969), *Chem. Rev.*, **69**, 838.
[116]   Hennion, G. F., McCusker, P. A., Ashby, E. C. and Rutkowski, A. J. (1957), *J. Am. Chem. Soc.*, **79**, 5194.
[117]   Brown, H. C. and Subba Rao, B. C. (1959), *J. Am. Chem. Soc.*, **81**, 6434; Brown, H. C. and Zweifel, G. (1966), *J. Am. Chem. Soc.*, **88**, 1433; (1967), **89**, 561.
[118]   Arnet, J. E. and Pettit, R. (1961), *J. Am. Chem. Soc.*, **83**, 2954; Rinehart, R. E. and Lasky, J. S. (1964), *J. Am. Chem. Soc.*, **86**, 2516.
[119]   Birch, A. J. and Subba Rao, G. S. R. (1968), *Tetrahedron Lett.*, 3797.
[120]   Alderson, T., Jenner, E. L. and Lindsey, R. V. (1965), *J. Am. Chem. Soc.*, **87**, 5638.
[121]   Lewandos, G. W. and Pettit, R. (1971), *Tetrahedron Lett.*, 789.
[122]   Hughes, W. B. (1972), *Organometal. Chem. Syn.*, **1**, 341.
[123]   Calderon, N. (1972), *Accts Chem. Res.*, **5**, 127.
[124]   Volger, H. C. (1967), *Rec. Trav. chim.*, **86**, 677.
[125]   van Helden, R. and Verberg, G. (1965), *Rec. Trav. chim.*, **84**, 1263.
[126]   Agnes, G., Chiusoli, G. P. and Cometti, G. (1970/1971), *Organometal. Chem. Syn.*, **1**, 93.
[127]   Brown, H. C., Hébert, N. C. and Snyder, C. H. (1961), *J. Am. Chem. Soc.*, **83**, 1001.
[128]   Brenner, W., Heimbach, P., Hey, H., Müller, E. W. and Wilke, G. (1969), *Liebigs Ann.*, **727**, 161.
[129]   Kiji, J., Masui, K. and Furukawa, J. (1970), *Chem. Comm.*, 1310; (1970), *Tetrahedron Lett.*, 2561.
[130]   Bird, C. W., Colinese, D. L., Cookson, R. C., Hudec, J. and Williams, R. O. (1961), *Tetrahedron Lett.*, 373.
[131]   Schrauzer, G. N., Ho, R. K. Y. and Schlesinger, G. (1970), *Tetrahedron Lett.*, 543.
[132]   Tsuji, J. (1973), *Accts Chem. Res.*, **6**, 8.
[133]   Suga, K., Watanabe, S. and Kamma, K. (1967), *Canad. J. Chem.*, **45**, 933.
[134]   Suga, K. and Watanabe, S. (1966), *J.C.S. Japan, Ind. Chem.*, **69**, 354.

[135]   Kurtz, P. (1962), *Liebigs Ann.*, **658**, 6.
[136]   Zweifel, G. and Miller, R. L. (1970), *J. Am. Chem. Soc.*, **92**, 6678.
[137]   Weigert, F. J., Baird, R. C. and Shapley, J. R. (1970), *J. Am. Chem. Soc.*, **92**, 6630.
[138]   Whitesides, G. M. and Ehmann, W. J. (1969), *J. Am. Chem. Soc.*, **91**, 3800.
[139]   Schäfer, W. (1966), *Angew. Chem. Internat. Edit.* **5**, 669; (1967), **6**, 518.
[140]   Whitesides, G. M. and Ehmann, W. J. (1970), *J. Am. Chem. Soc.*, **92**, 5625.
[141]   Wender, I., Sternberg, H. W. and Orchin, M. (1953), *J. Am. Chem. Soc.*, **75**, 3041.
[142]   Heck, R. F. and Breslow, D. S. (1961), *J. Am. Chem. Soc.*, **83**, 4023.
[143]   Casey, C. P. and Cyr, C. R. (1971), *J. Am. Chem. Soc.*, **93**, 1280.
[144]   Noack, K. and Calderazzo, F. (1967), *J. Organometal. Chem.*, **10**, 101.
[145]   Craig, P. J. and Green, M. (1967), *Chem. Comm.*, 1246.
[146]   Kubota, M. and Blake, D. M. (1971), *J. Am. Chem. Soc.*, **93**, 1368.
[147]   Casey, C. P. and Bunnell, C. A. (1971), *J. Am. Chem. Soc.*, **93**, 4077.
[148]   Brown, C. K. and Wilkinson, G. (1969), *Tetrahedron Lett.*, 1725.
[149]   Anon. (1971), *Chem. and Eng. News.*, **49**, 19.
[150]   Tsuji, J. and Ohno, K. (1966), *J. Am. Chem. Soc.*, **88**, 3452; 1968, **90**, 99.
[151]   Natta, G., Pino, P. and Ercoli, R. (1952), *J. Am. Chem. Soc.*, **74**, 4496.
[152]   Tsuji, J., Morikawa, M. and Kiji, J. (1964), *J. Am. Chem. Soc.*, **86**, 4851.
[153]   Tsuji, J., Kiji, J., Imamura, S. and Morikawa, M. (1964), *J. Am. Chem. Soc.*, **86**, 4350.
[154]   Yukawa, T. and Tsutsumi, S. (1969), *J. Org. Chem.*, **34**, 738.
[155]   Bittler, K., Kutepow, N., Neubauer, D. and Reis, H. (1968), *Angew. Chem. Internat. Edit.*, **7**, 329.
[156]   Tsuji, J., Kiji, J. and Morikawa, M. (1963), *Tetrahedron Lett.*, 1811.
[157]   Bird, C. W., Cookson, R. C., Hudec, J. and Williams, R. O. (1963), *J. Chem. Soc.*, 410.
[158]   Reppe, W. (1953), *Liebigs Ann.*, **582**, 1.
[159]   Seyferth, D. and Spohn, R. J. (1969), *J. Am. Chem. Soc.*, **91**, 3037.
[160]   Hirota, Y., Ryang, M. and Tsutsumi, S. (1971), *Tetrahedron Lett.*, 1531.
[161]   Bauld, N. L. (1963), *Tetrahedron Lett.*, 1841.
[162]   Corey, E. J. and Hegedus, L. S. (1969), *J. Am. Chem. Soc.*, **91**, 1233.
[163]   Cooke, M. P. (1970), *J. Am. Chem. Soc.*, **92**, 6080.
[164]   Collman, J. P., Winter, S. R. and Komoto, R. G. (1973), *J. Am. Chem. Soc.*, **95**, 249.
[165]   Collman, J. P., Winter, S. R. and Clark, D. R. (1972), *J. Am. Chem. Soc.*, **94**, 1788.
[166]   Ryang, M., Rhee, I. and Tsutsumi, S. (1964), *Bull. Chem. Soc., Japan*, **37**, 341.
[167]   Siegl, W. O. and Collman, J. P. (1972), *J. Am. Chem. Soc.*, **94**, 2516.
[168]   Brown, H. C. and Rathke, M. W. (1967), *J. Am. Chem. Soc.*, **89**, 2737, 2738.
[169]   Knights, E. F. and Brown, H. C. (1968), *J. Am. Chem. Soc.*, **90**, 5280.
[170]   Brown, H. C., Knights, E. F. and Coleman, R. A. (1969), *J. Am. Chem. Soc.*, **91**, 2144.
[171]   Brown, H. C. and Negishi, E. (1967), *J. Am. Chem. Soc.*, **89**, 5285.
[172]   Brown, H. C. and Pfaffenberger, C. D. (1967), *J. Am. Chem. Soc.*, **89**, 5475.
[173]   Brown, H. C. and Murray, K. (1959), *J. Am. Chem. Soc.*, **81**, 4108.
[174]   Coulson, D. R. (1969), *J. Am. Chem. Soc.*, **91**, 200.
[175]   Young, J. F., Osborn, J. A., Jardine, F. H. and Wilkinson, G. (1965), *Chem. Comm.*, 131.
[176]   O'Connor, C. and Wilkinson, G. (1969), *Tetrahedron Lett.*, 1375.
[177]   Meakin, P., Jesson, J. P. and Tolman, C. A. (1972). *J. Am. Chem. Soc.*, **94**, 3240.
[178]   Osborn, J. A., Jardine, F. H., Young, J. F. and Wilkinson, G. (1966), *J. Chem. Soc. (A)*, 1711.
[179]   Bhagwat, M. M. and Devaprabhakara, D. (1972), *Tetrahedron Lett.*, 1391.

[180]  Brown, M. and Piszkiewicz, L. W. (1967), *J. Org. Chem.*, **32**, 2013.
[181]  Harmon, R. E., Parsons, J. L., Cooke, D. W., Gupta, S. K. and Schoolenberg, J. (1969), *J. Org. Chem.*, **34**, 3684.
[182]  Birch, A. J. and Walker, K. A. M. (1966), *J. Chem. Soc. (C)*, 1894; (1967), *Tetrahedron Lett.*, 1935.
[183]  Ogata, I., Iwata, R. and Ikeda, Y. (1970), *Tetrahedron Lett.*, 3011; Hidai, M., Kuse, T., Hikita, T., Uchida, Y. and Misono, A. (1970), *Tetrahedron Lett.*, 1715.
[184]  Gilman, H. and Diehl, J. (1961), *J. Org. Chem.*, **26**, 4817.
[185]  Frainnet, E. and Esclamadon, C. (1962), *Compt. rend. Acad. Sci. Ser. C.*, **254**, 1814.
[186]  Kuivila, H. G. and Beumel, O. F. (1961), *J. Am. Chem. Soc.*, **83**, 1246.
[187]  Van Der Kerk, G. J. M., Noltes, J. G. and Suijten, J. G. H. (1937), *J. Appl. Chem.*, 7, 366.
[188]  Rothman, L. A. and Becker, E. I. (1960), *J. Org. Chem.*, **25**, 2203.
[189]  Kuivila, H. G. (1964), *Adv. Organometal. Chem.*, **1**, 47; (1968), *Accts Chem. Res.*, **1**, 289.
[190]  Alvernhe, G. and Laurent, A. (1972), *Tetrahedron Lett.*, 1007.
[191]  Berlin, K. D., Austin, T. H., Petersen, M. and Nagabushhnanam, M., (1964), *Topics in Phosphorus Chem.*, **1**, 17.
[192]  Seyfarth, H. E., Henkel, J. and Rieche, A. (1965), *Angew. Chem. Internat. Edit.*, **4**, 1074.
[193]  Lawesson, S. O. and Yang, N. C. (1959), *J. Am. Chem. Soc.*, **81**, 4230; Frisell, C. and Lawesson, S. O. (1961), *Org. Synth.*, **41**, 91.
[194]  Bordwell, F. G., Andersen, H. M. and Pitt, B. M. (1954), *J. Am. Chem. Soc.*, **76**, 1082; Trost, B. M. and Ziman, S. D. (1973), *J. Org. Chem.*, **38**, 932.
[195]  Zweifel, G. and Whitney, C. C. (1967), *J. Am. Chem. Soc.*, **89**, 2753.
[196]  Brown, H. C. and Subba Rao, B. C. (1959), *J. Am. Chem. Soc.*, **81**, 6423.
[197]  Brown, H. C. (1962), 'Hydroboration', Benjamin, New York; Zweifel, G. and Brown, H. C. (1963), *Org. Reactions*, **13**, 1.
[198]  Kuivila, H. G. and Amour, A. G. (1957), *J. Am. Chem. Soc.*, **79**, 5659.
[199]  Breuer, S. W., Leatham, M. J. and Thorpe, F. G. (1971), *Chem. Comm.*, 1475.
[200]  Brown, H. C., Heydkamp, W. R., Breuer, E. and Murphy, W. S. (1964), *J. Am. Chem. Soc.*, **86**, 3565.
[201]  Lane, C. F. and Brown, H. C. (1970), *J. Am. Chem. Soc.*, **92**, 7212.
[202]  Kuivila, H. G., Benjamin, L. E., Murphy, C. J., Price, A. D. and Polevy, J. H. (1962), *J. Org. Chem.*, **27**, 825.
[203]  Whitmore, F. C. and Hanson, E. R. (1941), *Org. Synth. Coll. Vol. 1*, 326.
[204]  Bernardi, A. (1926), *Gazz. chim. ital.*, **56**, 337.
[205]  Kitching, W. (1972), *Organometallic Reactions*, 3, 319.
[206]  Brown, H. C. and Geoghegan, P. J. (1967), *J. Am. Chem. Soc.*, **89**, 1522; (1970), *J. Org. Chem.*, **35**, 1844.
[207]  Pasto, D. J. and Gontarz, J. A. (1969), *J. Am. Chem. Soc.*, **91**, 719; Chambers, V. M. A., Jackson, W. R. and Young, G. W. (1971), *J. Chem. Soc. (C)*, 2075; Whitesides, G. M. and Filippo, J. S., (1970), *J. Am. Chem. Soc.*, **92**, 6611.
[208]  Brown, H. C. and Rei, M. H. (1969), *J. Am. Chem. Soc.*, **91**, 5646.
[209]  Brown, H. C., Geoghegan, P. J., Kurek, J. T. and Lynch, G. J. (1970), *Organometal. Chem. Syn.*, **1**, 7.
[210]  Bloodworth, A. J. and Bylina, G. S. (1972), *J.C.S. Perkin I*, 2433; Bloodworth, A. J. and Bunce, R. J. (1972), *J.C.S. Perkin I*, 2787.
[211]  Rappoport, Z., Winstein, S. and Young, W. G. (1972), *J. Am. Chem. Soc.*, **94**, 2320.
[212]  Foster, D. J. and Tobler, E. (1961), *J. Am. Chem. Soc.*, **83**, 851.
[213]  Sawai, H. and Takizawa, T. (1972), *Tetrahedron Lett.*, 4263.

[214]   Brown, H. C. and Kurek, J. T. (1969), *J. Am. Chem. Soc.*, **91**, 5647.
[215]   Aranda, V. G., Barluenga Mur, J., Asensio, G. and Yus, M. (1972), *Tetrahedron Lett.*, 3621.
[216]   Dobrev, A., Perie, J. J. and Lattes, A. (1972), *Tetrahedron Lett.*, 4013.
[217]   Heathcock, C. H. (1969), *Angew. Chem. Internat. Edit.*, **8**, 134.
[218]   Galle, J. E. and Hassner, A. (1972), *J. Am. Chem. Soc.*, **94**, 3930.
[219]   Curtin, D. Y. and Treten, J. L. (1961), *J. Org. Chem.,* **26**, 1764.
[220]   Pike, P. E., Marsh, P. G., Erickson, R. E. and Waters, W. L. (1970), *Tetrahedron Lett.*, 2679.
[221]   McKillop, A., Bromley, D. and Taylor, E. C. (1969), *Tetrahedron Lett.*, 1623; (1972), *J. Org. Chem.*, **37**, 88.
[222]   McKillop, A., Fowler, J. S., Zelesko, M. J., Hunt, J. D., Taylor, E. C. and McGillivray, G., (1969), *Tetrahedron Lett.*, 2423.
[223]   Taylor, E. C. and McKillop, A. (1970), *Accts Chem. Res.*, **3**, 338.
[224]   Taylor, E. C., Danforth, R. H. and McKillop, A. (1973), *J. Org. Chem.*, **38**, 2088.
[225]   Taylor, E. C., Altland, H. W., Danforth, R. H., McGillivray, G. and McKillop, A. (1970), *J. Am. Chem. Soc.*, **92**, 3520.
[226]   Chip, G. K., and Grossert, J. S. (1972), *J.C.S. Perkin I*, 1629.
[227]   Taylor, E. C., Kienzle, F. and McKillop, A. (1972), *Synthesis*, 38.
[228]   McKillop, A., Hunt, J. D., Zelesko, M. J., Fowler, J. S., Taylor, E. C., McGillivray, G. and Kienzle, F. (1971), *J. Am. Chem. Soc.*, **93**, 4841.
[229]   Ichikawa, K., Ikeda, Y. and Uemura, S. (1971), *Chem. Comm.*, 169.
[230]   McKillop, A., Hunt, J. D., Taylor, E. C. and Kienzle, F. (1970), *Tetrahedron Lett.*, 5275; McKillop, A., Hunt, J. D., Kienzle, F., Bigham, E. and Taylor, E. C. (1973), *J. Am. Chem. Soc.*, **95**, 3635.
[231]   McKillop, A., Swann, B. P. and Taylor, E. C. (1971), *J. Am. Chem. Soc.*, **93**, 4919; (1973), **95**, 3340.
[232]   McKillop, A., Oldenziel, O. H., Swann, B. P., Taylor, E. C. and Robey, R. L. (1971), *J. Am. Chem. Soc.*, **93**, 7333.
[233]   Kruse, W. and Bednarski, T. M. (1971), *J. Org. Chem.*, **36**, 1154.
[234]   Henry, P. M. (1971), *J. Org. Chem.*, **36**, 1886.
[235]   Bushweller, C. H. (1968), *Tetrahedron Lett.*, 6123.
[236]   Smidt, J. (1962), *Chemy Ind.*, 54.
[237]   Davies, N. R. (1967), *Rev. Pure Appl. Chem.*, **17**, 83.
[238]   Stern, E. W. and Spector, M. C. (1961), *Proc. Chem. Soc.* 370.
[239]   Baird, W. C. (1966), *J. Org. Chem.*, **31**, 2411.
[240]   Kitching, W., Rappoport, Z., Winstein, S. and Young, W. G. (1966), *J. Am. Chem. Soc.*, **88**, 2054.
[241]   Henry, P. M. (1973), *Accts Chem. Res.*, **6**, 16.
[242]   Henry, P. M. (1972), *J. Org. Chem.*, **37**, 7311.
[243]   Brady, D. G. (1970), *Chem. Comm.*, 434.
[244]   Lloyd, W. G. and Luberoff, B. J. (1969), *J. Org. Chem.*, **34**, 3949.
[245]   Palumbo, R., De Renzi, A., Panunzi, A. and Paiaro, G. (1969), *J. Am. Chem. Soc.*, **91**, 3874; Panunzi, A., De Renzi, A. and Paiaro, G. (1970), *J. Am. Chem. Soc.*, **92**, 3488.
[246]   Stern, E. W. and Spector, M. L. (1966), *J. Org. Chem.*, **31**, 596.
[247]   Rosenthal, A., and Wender, I. (1968), in 'Organic Syntheses via Metal Carbonyls', Wender I. and Pino, P., Ed., Interscience, New York, p. 405.
[248]   Angelici, R. J. (1972), *Accts Chem. Res.*, **5**, 335.
[249]   Fukuoka, S., Ryang, M., and Tsutsumi, S. (1971), *J. Org. Chem.*, **36**, 2721.
[250]   Blum, J. and Scharf, G. (1970), *J. Org. Chem.*, **35**, 1895.

145

# Index